FinFET Modeling for IC Simulation and Design

Using the BSIM-CMG Standard

Yogesh Singh Chauhan

Darsen D. Lu

Sriramkumar Vanugopalan

Sourabh Khandelwal

Juan Pablo Duarte

Navid Paydavosi

Ai Niknejad

Chenming Hu

AMSTERDAM · BOSTON · HEIDELBERG · LONDON
NEW YORK · OXFORD · PARIS · SAN DIEGO
SAN FRANCISCO · SINGAPORE · SYDNEY · TOKYO
Academic Press is an imprint of Elsevier

ELSEVIER

Academic Press is an imprint of Elsevier
125 London Wall, London, EC2Y 5AS, UK
525 B Street, Suite 1800, San Diego, CA 92101-4495, USA
225 Wyman Street, Waltham, MA 02451, USA
The Boulevard, Langford Lane, Kidlington, Oxford OX5 1GB, UK

Notices
Knowledge and best practice in this field are constantly changing. As new research and
experience broaden our understanding, changes in research methods, professional practices,
or medical treatment may become necessary.

Practitioners and researchers must always rely on their own experience and knowledge in
evaluating and using any information, methods, compounds, or experiments described herein.
In using such information or methods they should be mindful of their own safety and the
safety of others, including parties for whom they have a professional responsibility.

To the fullest extent of the law, neither the Publisher nor the authors, contributors, or editors,
assume any liability for any injury and/or damage to persons or property as a matter of
products liability, negligence or otherwise, or from any use or operation of any methods,
products, instructions, or ideas contained in the material herein.

Library of Congress Cataloging-in-Publication Data
A catalog record for this book is available from the Library of Congress

British Library Cataloguing in Publication Data
A catalogue record for this book is available from the British Library

For information on all Academic Press publications
visit our web site at store.elsevier.com

Printed and bound in the USA

ISBN: 978-0-12-420031-9

Contents

Author Biographies... ix
Preface xi

CHAPTER 1 FinFET—From device concept to standard compact model 1
   1.1 The root cause of short-channel effects in the twenty-first century MOSFETs 2
   1.2 The thin-body MOSFET concept................................... 4
   1.3 The FinFET and a new scaling path for MOSFETs 4
   1.4 Ultra-thin-body FET 6
   1.5 FinFET compact model—the bridge between FinFET technology and IC design... 7
   1.6 A brief history of the first standard compact model, BSIM 8
   1.7 Core and real-device models....................................... 9
   1.8 The industry standard FinFET compact model 11
   References 12

CHAPTER 2 Compact models for analog and RF applications 15
   2.1 Introduction... 15
   2.2 Important compact model metrics 16
   2.3 Analog metrics 16
      2.3.1 Quiescent operating point............................... 17
      2.3.2 Geometric scalability 19
      2.3.3 Variability model 23
      2.3.4 Intrinsic voltage gain................................... 24
      2.3.5 Speed: Unity gain frequency........................... 31
      2.3.6 Noise 32
      2.3.7 Linearity and symmetry................................ 36
      2.3.8 Symmetry 42
   2.4 RF metrics 44
      2.4.1 Two-port parameters 44
      2.4.2 The need for speed 46
      2.4.3 Non-quasi-static model 55
      2.4.4 Noise 58
      2.4.5 Linearity 64
   2.5 Conclusion... 68
   References 68

CHAPTER 3 Core model for FinFETs 71
   3.1 Core model for double-gate FinFETs 72
   3.2 Unified FinFET compact model 80

Chapter 3 Appendix: Explicit surface potential model 87
3A.1 Continuous starting function 88
3A.2 Quartic modified iteration: Implementation and evaluation 91
References 96

CHAPTER 4 Channel current and real device effects **99**
4.1 Introduction 99
4.2 Threshold voltage roll-off 100
4.3 Subthreshold slope degradation 106
4.4 Quantum mechanical v_{th} correction 107
4.5 Vertical-field mobility degradation 109
4.6 Drain saturation voltage, v_{dsat} 109
　　4.6.1　Extrinsic case (RDSMOD = 1 and 2) 110
　　4.6.2　Intrinsic case (RDSMOD = 0) 111
4.7 Velocity saturation model 114
4.8 Quantum mechanical effects 115
　　4.8.1　Effective width model 118
　　4.8.2　Effective oxide thickness/effective capacitance 118
　　4.8.3　Charge centroid calculation for accumulation 119
4.9 Lateral nonuniform doping model 119
4.10 Body effect model for a bulk FinFET (BULKMOD = 1) 119
4.11 Output resistance model 120
　　4.11.1　Channel-length modulation 121
　　4.11.2　Drain-induced barrier lowering 122
4.12 Channel current 123
References 124

CHAPTER 5 Leakage currents **127**
5.1 Weak-inversion current 129
5.2 Gate-induced source and drain leakages 130
　　5.2.1　GIDL/GISL current formulation in BSIM-CMG 132
5.3 Gate oxide tunneling 133
　　5.3.1　Gate oxide tunneling formulation in BSIM-CMG 134
　　5.3.2　Gate-to-body tunneling current in
　　　　　　depletion/inversion 135
　　5.3.3　Gate-to-body tunneling current in accumulation 136
　　5.3.4　Gate-to-channel tunneling current in inversion 137
　　5.3.5　Gate-to-source/drain tunneling current 138
5.4 Impact ionization 140
References 141

**CHAPTER 6 Charge, capacitance, and non-quasi-static
effects** **143**
6.1 Terminal charges 144
　　6.1.1　Gate charge 144
　　6.1.2　Drain charge 145

6.1.3 Source charge... 146
6.2 Transcapacitances 146
6.3 Non-quasi-static effects models 147
6.3.1 Relaxation time approximation model 149
6.3.2 Channel-induced gate resistance model 151
6.3.3 Charge segmentation model 152
References 155

CHAPTER 7 Parasitic resistances and capacitances **157**
7.1 FinFET device structure and symbol definitions 158
7.2 Modeling of geometry-dependent source/drain resistances
in FinFETs.. 161
7.2.1 Contact resistance 162
7.2.2 Spreading resistance 164
7.2.3 Extension resistance.................................... 167
7.3 Parasitic resistance model verification 169
7.3.1 TCAD simulation setup................................. 169
7.3.2 Device optimization..................................... 170
7.3.3 Extraction of source and drain resistances 172
7.3.4 Discussion 176
7.4 Implementation considerations of the parasitic
resistance model.. 178
7.4.1 Physical parameters 178
7.4.2 Resistance components 178
7.5 Gate electrode resistance model 179
7.6 FinFET parasitic capacitance models 179
7.6.1 Connection of parasitic capacitance components....... 179
7.6.2 Derivation of two-dimensional fringe capacitance 181
7.7 Modeling of FinFET fringe capacitance in three dimensions:
CGEOMOD = 2 187
7.8 Parasitic capacitance model verification 188
7.9 Summary 192
References 193

CHAPTER 8 Noise **195**
8.1 Introduction.. 195
8.2 Thermal noise 196
8.3 Flicker noise 198
8.4 Other noise components 201
8.5 Summary 201
References 201

CHAPTER 9 Junction diode *I-V* and *C-V* models **203**
9.1 Junction diode current model 205
9.1.1 Reverse-bias additional leakage model 208

9.2 Junction diode charge/capacitance model 210

    9.2.1 Reverse-bias model 210

    9.2.2 Forward-bias model 213

References 216

CHAPTER 10 Benchmark tests for compact models **217**

    10.1 Asymptotic correctness 218

    10.2 Benchmark tests 219

        10.2.1 Tests for checking physical behavior in
weak-inversion and strong-inversion regions 219

        10.2.2 Symmetry tests 222

        10.2.3 Reciprocity test for capacitances in a compact
model 226

        10.2.4 Test for the self-heating effect model 227

        10.2.5 Tests for the thermal noise model 227

    References 228

CHAPTER 11 BSIM-CMG model parameter extraction **231**

    11.1 Parameter extraction background 232

    11.2 BSIM-CMG parameter extraction strategy 233

    11.3 Conclusion 242

    References 242

CHAPTER 12 Temperature dependence **245**

    12.1 Semiconductor properties 246

        12.1.1 Band gap temperature dependence 246

        12.1.2 Temperature dependence of N_c, v_{bi}, and Φ_b 246

        12.1.3 Temperature dependence of the intrinsic carrier
concentration 247

    12.2 Temperature dependence of the threshold voltage 247

        12.2.1 Temperature dependence of drain-induced barrier
lowering 248

        12.2.2 Temperature dependence of the body effect 248

        12.2.3 Subthreshold swing 248

    12.3 Temperature dependence of mobility 249

    12.4 Temperature dependence of velocity saturation 249

        12.4.1 Temperature dependence of the nonsaturation
effect 250

    12.5 Temperature dependence of leakage currents 250

        12.5.1 Gate current 250

        12.5.2 Gate-induced drain/source leakage 250

        12.5.3 Impact ionization 251

    12.6 Temperature dependence of parasitic source/drain
resistances 251

12.7 Temperature dependence of source/drain diode
       characteristics 252
       12.7.1 Direct current model 252
       12.7.2 Capacitance.. 254
       12.7.3 Trap-assisted tunneling current 254
12.8 Self-heating effect.. 256
12.9 Validation range 257
12.10 Model validation on measured data 257
References 260

Appendix... 261
Index 287

Author Biographies

Yogesh S. Chauhan is an Assistant Professor in the electrical engineering department at the Indian Institute of Technology, Kanpur. He received his Ph.D. in compact modeling of high voltage MOSFETs in 2007 from EPFL, Switzerland. From 2007-2010, he was a manager at IBM, Bangalore, where he led a compact modeling team, focusing on RF bulk and SOI transistors and ESD modeling. From 2010-2012, he was a postdoctoral fellow at the University of California, Berkeley, where he worked on the development of bulk and multigate transistor models, including BSIM6, BSIM-IMG and BSIM-CMG. He received the IBM Faculty Award in 2013 for his contribution to compact modeling. He has co-authored over 50 conference and journal publications in the field of device compact modeling.

Darsen D. Lu was one of the key contributors of the industry standard FinFET compact model, BSIM-CMG, and thin-body SOI compact model, BSIM-IMG. He received his B.Sc. in electrical engineering in 2005, from National Tsing Hua University, Hsinchu, Taiwan, and his M.Sc. and Ph.D. in electrical engineering from the University of California, Berkeley, in 2007 and 2011 respectively. Since 2011, he has been a research scientist at the IBM Thomas J. Watson Research Center, Yorktown Heights, New York. His current research focuses on the modeling of novel semiconductor devices such as SiGe FinFETs, phase change memory and carbon-based transistors.

Sriramkumar Venugopalan received his M.Sc. and Ph.D. in electrical engineering at the University of California, Berkeley and his B.Sc. from the Indian Institute of Technology (IIT), Kanpur. While at Berkeley he worked in the BSIM Group and pursued research and development of multi-gate transistor compact SPICE models that contributed to the industry standard BSIM-CMG model. He has authored and co-authored more than 30 research papers in the area of semiconductor device SPICE models and integrated circuit design. Currently Dr. Venugopalan is with Samsung Electronics pursuing RF integrated circuit design in advanced semiconductor technology nodes.

Sourabh Khandelwal is currently a Postdoctoral Researcher in the BSIM Group, University of California, Berkeley. He received his Ph.D. from the Norwegian University of Science and Technology in 2013 and his M.Sc. from the Indian Institute of Technology (IIT), Bombay in 2007. From 2007–2010 he worked as a research engineer at the IBM Semiconductor Research and Development Centre, developing compact models for RF SOI devices. He holds a patent and has authored several research papers in the area of device modeling and characterization. His Ph.D. work on the GaN compact model is under consideration for industry standardization by the Compact Model Coalition.

Juan Pablo Duarte Sepúlveda is currently working towards his Ph.D. at the University of California, Berkeley. He received his B.Sc. in 2010 and his M.Sc. in 2012, both in electrical engineering from the Korea Advanced Institute of Science and Technology (KAIST). He held a position as a lecturer at the Universidad Tecnica Federico Santa Maria, Valparaiso, Chile, in 2012. He has authored many papers on nanoscale semiconductor device modeling and characterization. He received the Best Student Paper Award at the 2013 International Conference on Simulation of Semiconductor Processes and Devices (SISPAD) for the paper: Unified FinFET Compact Model: Modelling Trapezoidal Triple-Gate FinFETs.

Navid Paydavosi received his Ph.D. in Micro-Electro-Mechanical Systems (MEMS) and Nanosystems from the University of Alberta, Canada in 2011. He worked for the BSIM Group at the University of California, Berkeley, as a post-doctoral scholar from 2012 to 2014. He has published several research papers on the theory and modeling of modern Si-MOSFETs and there future alternatives, including carbon-based and III-V high electron mobility devices. Currently Dr. Paydavosi is with Intel Corporation, Oregon as a device engineer working on process technology development.

Ali M. Niknejad received his B.Sc in electrical engineering from the University of California, Los Angeles, in 1994, and his M.Sc. and Ph.D., also in electrical engineering, from the University of California, Berkeley, in 1997 and 2000 respectively. He is currently a professor in the EECS department at UC Berkeley and Faculty Director of the Berkeley Wireless Research Center (BWRC) Group. Professor Niknejad was the recipient of the 2012 ASEE Frederick Emmons Terman Award for his work and textbook on electromagnetics and RF integrated circuits. He has co-authored over 200 conference and journal publications in the field of integrated circuits and device compact modeling. His focus areas of research include analog, RF, mixed-signal, mm-wave circuits, device physics and compact modeling, and numerical techniques in electromagnetics.

Chenming Hu is Distinguished Chair, Professor Emeritus at the University of California, Berkeley. He was the Chief Technology Officer of TSMC and founder of Celestry Design Technologies. He is best known for developing the revolutionary 3D transistor FinFET that powers semiconductor chips beyond 20nm. He also led the development of BSIM - the industry standard transistor model that is used in designing most of the integrated circuits in the world. He is a member of the US Academy of Engineering, the Chinese Academy of Science, and Academia Sinica. His honors include the Asian American Engineer of the Year Award, the IEEE Andrew Grove Award and Solid Circuits Award as well as the Nishizawa Medal, and UC Berkeley's highest honor for teaching - the Berkeley Distinguished Teaching Award.

Preface

If you have opened this book, you probably know that the first 3D transistor, the FinFET, and its adoption by industry for its power and speed advantages have been the biggest semiconductor news in recent years. The FinFET has been called the most drastic shift in semiconductor technology in over 40 years.

Because a book reflects the background of its authors, it serves the readers for us to describe our background. We are two professors and six current or former Ph.D. students or postdoctoral researchers of the University of California, Berkeley. One of the professors (C.H.) was the lead inventor and developer of the FinFET. The other professor (A.N.) pioneered the field of 100 GHz CMOS. All the coauthors are or were members of the BSIM research group that created the industry-standard FinFET model for simulation and design of FinFET-based ICs. BSIM planar CMOS models have been used to design IC products with estimated cumulative sales of a trillion US dollars since 1997. One may expect the BSIM FinFET model to have a similarly large impact in the future.

We wrote this book for IC designers, device engineers, researchers, and students. It presents the what, why, and how of the FinFET and compact modeling; models for analog and RF applications; and a thorough discussion of the BSIM FinFET model (BSIM-CMG). We start from the ABC of the FinFET and end with the XYZ of the FinFET model. Even if you are familiar with BSIM-CMG, you may be surprised to learn that it can model FinFETs with arbitrary fin shapes such as trapezoidal, round-corner, cylindrical, and even asymmetrical shapes. It also models FinFETs employing non-silicon channel materials such as SiGe, Ge, and InGaAs. You are holding the best handbook for the FinFET model for IC simulation.

We do not claim to have created the equations and model presented in this book. Many other former members of the BSIM group contributed to the creation of BISM-CMG. We acknowledge their indirect contributions to this book. Most notably, Chung-Hsun Lin and Mohan Dunga were the first student developers of BSIM-CMG, starting in 2004. Other direct or indirect contributors to the book include Walter Li, Wei-Man Lin, Shijin Yao, Muhammed Karim, Chandan Yadav, and Avirup Dasgupta.

We thank the many industry BSIM users who helped to make BSIM-CMG a better model for their corporate employers. They did so by testing the beta model and pointing out its weaknesses in accuracy or robustness during the 2-year-long evaluation and (uncontested) election process of the standard FinFET model. The list includes R. Williams (IBM); A.S. Roy, S. Mudanai (Intel); K.-W. Su, W.-K. Lee, M.-C. Jeng (Taiwan Semiconductor Manufacturing Company); J.-S. Goo (Globalfoundries); P. Lee (Micron Technology); Q. Wang, J. Wang, W. Liu

(Synopsys); J. Xie, F. Zhao (Cadence); A. Ramadan, S. Mohamed, A.-E. Ahmed (Mentor Graphics); P. O'Halloran (Tiburon Design Automation); B. Chen, S. Mertens (Accelicon/Agilent, now Keysight Technologies); J. Ma (ProPlus); and G. Coram (Analog Devices).

Most importantly, we wish to express our deepest gratitude to our families, who tolerated and made tolerable our long hours in the office and at computers.

And we thank you, dear readers, for giving our book meaning by using it.

The authors

FinFET—From device concept to standard compact model

1

CHAPTER OUTLINE

1.1 The root cause of short-channel effects in the twenty-first century MOSFETs 2
1.2 The thin-body MOSFET concept 4
1.3 The FinFET and a new scaling path for MOSFETs 4
1.4 Ultra-thin-body FET 6
1.5 FinFET compact model—the bridge between FinFET technology and IC design 7
1.6 A brief history of the first standard compact model, BSIM 8
1.7 Core and real-device models 9
1.8 The industry standard FinFET compact model..11
References...12

Part of the semiconductor industry's formula for success is to make incremental changes, not drastic changes. The planar MOSFET has served the electronics industry well for 40 years. Aggressive engineering has managed to reduce its size again and again without change to its basic structure. Yet the IC design window for performance, power consumption, and sensitivity to device variation has shrunk to the point that a major change to a better transistor structure is unavoidable. The FinFET is that new better transistor. Adopting the FinFET has been called the most drastic shift in semiconductor technology in over 40 years. The FinFET provides relief from the performance, power, and device variation predicaments that the IC industry has struggled with in the past decade. More importantly, it redirects device scaling from a path heading for the cliff to a new one that allows scaling to continue onward as long as lithography allows.

A brand new transistor requires a new design infrastructure to enable the design of FinFET-based circuits and products. The foundation of the infrastructure is a computationally very efficient mathematical model that nevertheless represents a FinFET very accurately. It is called a compact model, or a SPICE model.

This chapter presents the FinFET—what it is, what it does, and what new scaling concept inspired its invention. This chapter also introduces the role of a compact model in the semiconductor industry and the industry's first and dominant standard compact model, BSIM. The BSIM FinFET model is enabling the design of new generations of ICs with higher performance, lower power consumption, and higher layout density.

1.1 THE ROOT CAUSE OF SHORT-CHANNEL EFFECTS IN THE TWENTY-FIRST CENTURY MOSFETs

What is the basic concept behind this remarkable FinFET? In order to understand the pain reliever, we must first understand the pain itself. As the gate length shrinks, the MOSFET's I_d-V_g characteristics degrade in two major ways as illustrated in Figure 1.1. First, the subthreshold swing (S) degrades and V_t decreases—that is, the device cannot be turned off easily by lowering V_g. Second, S and V_t become increasingly sensitive to L_g variations—that is, device variations become more problematic. These problems are known as short-channel effects.

Figure 1.2 shows the root cause of short-channel effects in the twentieth century MOSFETs [1]. A transistor is turned on and off when V_g lowers and raises the potential of the channel (and thus the potential barrier between the channel and the source) through the gate-to-channel capacitance, C_g. In an ideal transistor, the channel potential is controlled only by V_g and C_g. In a real transistor, the channel potential is also subjected to the influence of V_d through C_d. When L_g is large, C_d is much smaller than C_g and the drain voltage does not interfere with V_g's role as the sole controlling voltage. As L_g decreases, C_d increases [1, 2] and V_g loses its absolute control. In extreme cases, V_g has less control than V_d and the transistor

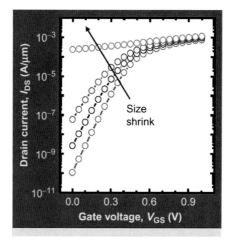

FIGURE 1.1

Shrinking gate length makes the sub-V_t swing larger, and V_t and I_{off} more sensitive to gate critical dimension variation and random dopant fluctuation.

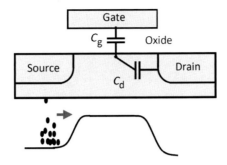

FIGURE 1.2

With decreasing L, rising C_d allows V_d to control the channel potential (bottom) just as V_g [1]. Scaling the gate oxide thickness was a good solution for this problem in the twentieth century.

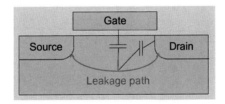

FIGURE 1.3

Beyond 20 nm even a 0 nm thick gate dielectric cannot stop V_d from lowering the potential barrier along leakage paths a few nanometers below the interface [1].

can be turned on by V_d alone without V_g as shown in the top curve in Figure 1.1. Before that extreme is reached, we get the other deteriorating curves in Figure 1.1. The solution that worked well in the twentieth century was to increase C_g by reducing the gate oxide thickness in proportion to L_g.

However, there is a new root cause that would limit L_g scaling even if an ideal "zero thickness" dielectric were available. Figure 1.3 shows that the leakage current does not have to flow along the silicon-dielectric interface. Leakage paths far from the gate as illustrated in Figure 1.3 are worse than the surface leakage path because they are only weakly controlled by V_g—that is, "C_g" is small even with zero oxide thickness. As a result, the potential barriers along these weakly controlled paths can be easily lowered by V_d through the large C_d in a small-L_g device.

Having identified this new root cause, researchers at the University of California, Berkeley proposed a new thin-body scaling concept and the FinFET structure to the US government's Defense Advanced Research Project Agency in response to a request for the proposal of sub-25-nm switching devices in 1996.

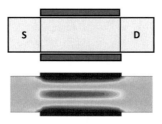

FIGURE 1.4

A thin silicon body eliminates the leakage paths in Figure 1.3 (top) and leakage current density is low near the gate and highest in the center of the body (bottom) [1].

1.2 THE THIN-BODY MOSFET CONCEPT

Figure 1.4 shows a MOSFET whose body is a thin piece of silicon with gates above and below it. If the body is thin, any lines drawn between the source and the drain (any potential leakage paths) would not be far from one or the other gate. The thin-body concept eliminates the need for heavy channel doping for suppressing the short-channel effects. Channel doping may still be used to adjust the threshold voltage in the near term, but that function can be performed by gate metal work function engineering in order to realize the full potential of future thin-body transistors. Random dopant fluctuation, a major and fundamental contributor to device variation, can be eliminated. Channel carrier mobility and junction leakage are improved. An undoped body reduces the electric field normal to the semiconductor and oxide interface. This should improve the temperature bias instability (negative-bias temperature instability and positive-bias temperature instability) and the gate dielectric tunneling leakage and wear out.

1.3 THE FinFET AND A NEW SCALING PATH FOR MOSFETs

The FinFET in Figure 1.5 is a manufacturable version of the thin-body transistor in Figure 1.4. The thin body is shaped like a fish fin and is created with the usual patterning and etching technologies. The fin can be constructed on silicon-on-insulator (SOI) or lower-cost bulk substrates. The fin is then processed into a FinFET in very much the same way as for processing of a planar MOSFET because the fin is shorter than the gate thickness, so the structure is quasi-planar.

A FinFET is dense and manufacturable. Figure 1.6 shows that the only major new fabrication step is overetching of the shallow trench isolation oxide. A FinFET occupies less silicon area than a planar MOSFET because the channel width (W) of a FinFET [3] is the peripheral length of the fin that includes all sides of the fin cross-section, and W can be significantly larger than the fin pitch. The multiple fin height [4] in Figure 1.6 can improve the density of SRAM. In the future, a fin height increase may supplement lithography shrinkage.

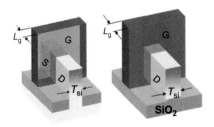

FIGURE 1.5

A FinFET with a thin body ($T_{si} < L_g$) dramatically reduces short-channel effects, including device variation and I_{off}.

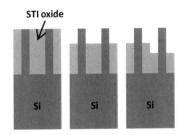

FIGURE 1.6

After shallow trench isolation (STI) chemical-mechanical polishing (left), overetching of oxide exposes the fins of the FinFETs (middle). Multiple fin heights (right) can improve circuit density [5].

Intel was the first company to introduce FinFET technology into mass production, in 2011, and reported that the manufacturing cost is comparable to that of the conventional planar MOSFET technology, higher by only a low single-digit percent. It also reported that no special manufacturing equipment is required.

As long as the fin thickness (body thickness), T_{si}, is smaller than L_g, the short-channel effects are well suppressed and the subthreshold swing is basically the theoretical best case, approximately 62 mV per decade at room temperature (Figure 1.7). A new scaling path was born: L_g can be scaled by scaling the fin (body) thickness. If lithography and etching can produce 5 nm L_g, for example, they can produce approximately 5 nm T_{si}. Therefore, the condition $T_{si} \lesssim L_g$ can always be satisfied. In 1999, 18 and 45 nm working FinFETs and 10 nm FinFET simulation results were reported [3]. Soon after, 10 and then 5 nm FinFETs [5] were reported by IC manufacturers.

Theoretically and practically, FinFET technology is scalable to single-digit nanometers. FinFET technology can propel ICs to the end of lithography scaling. In addition to scaling, researchers are also making FinFETs with group III-V materials and germanium [6, 7]. Germanium and group III-V materials such as InGaAs have intrinsically much higher carrier mobility than silicon. This can improve the device performance significantly.

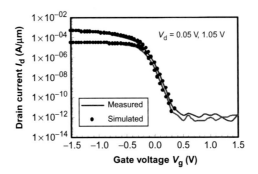

FIGURE 1.7

A 1999 FinFFT with excellent subthreshold swing and I_{off} and undoped body eliminating the random dopant effect. L_g = 45 nm. T_{si} = 30 nm [3].

1.4 ULTRA-THIN-BODY FET

In 1996, we proposed two implementations of the thin-body concept. They are the FinFET and ultrathin body (UTB). While the FinFET is now the mainstream advanced technology, examining a FinFET together with UTB puts the FinFET in perspective. A FinFET's ability to suppress the short-channel effects does not arise from it being three-dimensional, although that gives the FinFET an advantage in layout density. It arises from having a thin body that precludes the presence of a semiconductor (potential leakage paths) that is not very close to the gate.

Reducing the silicon film thickness or silicon doping concentration from partially to fully deleted SOI (e.g., from 40 to 15 nm) would not improve the short-channel effects and may worsen them by eliminating the ground plane effect provided by the undepleted silicon body [8]. However, if the silicon film is only several nanometers thick as shown in Figure 1.8, short-channel effects can be greatly suppressed [9] by

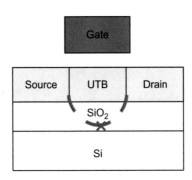

FIGURE 1.8

Ultrathin body (UTB) has no vulnerable leakage paths if silicon exists only within a few nanometers from the gate [7].

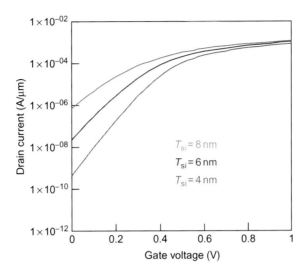

FIGURE 1.9

Simulation shows UTB SOI has excellent I_{off} and S if $T_{si} < L_g/4$. $L_g = 20$ nm, $V_{ds} = 1$ V, lightly doped body [7].

eliminating the worst leakage paths. Figure 1.9 illustrates the compelling benefit of UTB: simulated leakage current is reduced by approximately 10 times for every 1 nm drop of the body thickness—in the UTB regime.

The current implementation of UTB requires SOI substrates with silicon film uniformity of ± 0.5 nm, or less than two silicon atoms, so that a 5 nm ultrathin silicon film will not have more than $\pm 10\%$ nonuniformity. In the future, two-dimensional semiconductors such as graphene, WSe_2, or MoS_2 may provide the ultimate single-molecule thin body for UTB transistors.

1.5 FinFET COMPACT MODEL—THE BRIDGE BETWEEN FinFET TECHNOLOGY AND IC DESIGN

Designing FinFET-based ICs requires a FinFET model for circuit simulation. In order to design ICs, design teams need two things from their foundry partners or the wafer manufacturing divisions of their companies: design rules and the SPICE model. The SPICE model is also known as a compact model. Design rules are created by each manufacturing company. The SPICE model, which is a set of long equations that are capable of reproducing the very complex transistor characteristics accurately and fast for SPICE simulation, is likely a free industry standard model provided by a university.

The compact model is effectively a "contract" between the wafer fab and the IC designers. "First wafer success" has become the norm in the past dozen years in spite

Compact model is the

Major bridge for
information exchange

Device/Fab technology　　　Circuit/product design

Example: BSIM
- **First industry standard compact model**
- ~25,000 lines of C code ~8000 lines of Verilog-A

- Accurate: **Fit MOSFET data with ~1% RMS error**
- Fast: ~10 μs per bias point
- Smooth

FIGURE 1.10

Complex device and manufacturing technologies determine the electrical behaviors of a transistor. A compact model captures this information and makes it available for IC circuit and product design.

of the rising technology complexity and circuit size and variety. That is basically due to the improved accuracy of the compact models. If the model does not describe the transistor characteristics accurately, no amount of design effort can guarantee design success. The model equations contain adjustable parameters. Engineers, with the help of automated parameter optimization tools, choose the parameter values so that the model accurately reproduces the current, capacitance, and noise over many orders of magnitude for the entire operating range of terminal voltages, gate lengths and widths, and temperatures. The accuracy of the final result greatly depends on the model equations. The compact model may be used with SPICE simulators to simulate and design circuits directly. Or it may be used with "fast SPICE," which achieves order-of-magnitude speedup over SPICE with some loss of accuracy.

A good compact model must be very accurate to avoid expensive design respins, very fast to support simulation of large circuits, and very robust for convergence in a wide range of complex circuits (Figure 1.10).

1.6 A BRIEF HISTORY OF THE FIRST STANDARD COMPACT MODEL, BSIM

BSIM stands for "Berkeley short-channel insulated-gate field effect transistor model." "Insulated-gate field effect transistor" is an old name for MOSFET. BSIM was the first industry standard model and continues to be the most popular compact model today. BSIM's genesis may be traced to BSIM1 published in 1984 [10], which was followed by BSIM2 in 1988 [11].

At the same time, the Berkeley research team had a very productive parallel research effort in MOSFET physics and technology. Our research into the many devices' physics effects and behaviors of aggressively scaled MOSFETs gradually built a collection of models for V_t dependence of bias and gate length, mobility degradation, velocity saturation effect, output conductance, unified flicker noise theory, etc. Eventually, these models became the building blocks for new versions of BSIM models.

BSIM3 incorporated some of the new original device physics models. That approach was a marked departure from all previous compact models, including BSIM1 and BSIM2. Those models used simplistic device physics and relied heavily on "curve fitting." That approach may be acceptable for modeling I_d versus V_d, for example, but is inadequate for modeling the transconductance and the higher-order derivatives.

BSIM3 [12] was such an improvement over the previous models that the Compact Model Council, an industry standard organization formed in 1995, selected BSIM3v3 as the world's first industry standard model. Soon after, BSIM3 replaced many dozens of SPICE models in use in 1995. BSIM has been provided by the University of California, Berkeley to users worldwide royalty free following the tradition of the Berkeley SPICE.

1.7 CORE AND REAL-DEVICE MODELS

All compact MOSFET models start with a "core model" that models a prototype very long-channel transistor. For the other 99.99% of the transistors used in an IC, the accuracy is achieved with numerous add-on "real-device models" as shown in Figure 1.11 . With the CMOS technology aggressively scaled, the real-device effects have become the dominant, not the secondary, effects, and the real-device models determine the accuracy of circuit simulation. BSIM excels because of its accurate real-device models (see Chapter 4). For example, the output resistance used to be modeled with an empirical constant early voltage model. BSIM3 [10] introduced three separate physical mechanisms—channel length modulation, drain-induced barrier lowering, and hot-carrier-induced body bias effect. Each of these three mechanisms is modeled with a nonlinear multivariable function of channel length oxide thickness, V_t, V_{ds}, V_{gs}, and V_{bs}. An accurate output resistance model is very important for analog circuit design, and the BSIM output conductance model was an instant success and continues to be used today.

Another example is the gate-induced drain leakage (GIDL). It was introduced into BSIM3 after we had discovered this new leakage current and explained it as the band-to-band tunneling current induced by the gate-to-drain voltage [13]. Once the mechanism was clearly understood, a simple analytical model could be developed, and it proved to be very accurate for all subsequent generations of MOSFET technology.

Yet another example is the flicker noise, or $1/f$ noise. It unified the noises due to fluctuation in the number of the channel inversion charge carriers and the fluctuation in the coulombic scattering mobility. They can be unified because both result from

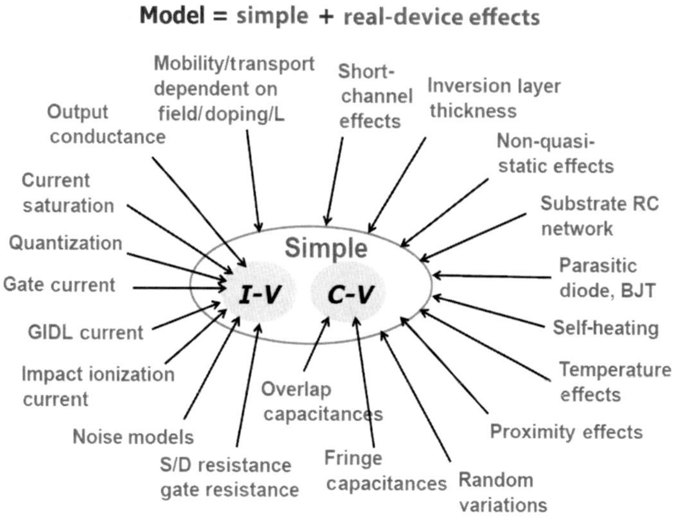

Model = simple + real-device effects

Mobility/transport dependent on field/doping/L

Short-channel effects

Inversion layer thickness

Output conductance

Current saturation

Quantization

Gate current

GIDL current

Impact ionization current

Noise models

Non-quasi-static effects

Substrate RC network

Parasitic diode, BJT

Self-heating

Temperature effects

Proximity effects

Simple

I-V   C-V

Overlap capacitances

S/D resistance gate resistance

Fringe capacitances

Random variations

FIGURE 1.11

A compact model is formed from a simple (long-channel) model and numerous real-device models. The latter constitute 90% of the model and are responsible for the global accuracy.

the capture and emission of electrons or holes by the charge traps in SiO_2 near the interface [14]. We validated the model in detail using the random telegraphic noise measurements that can only be observed in transistors with such small length and width that one transistor contains only one or two observable oxide traps. These physics studies led to an accurate BSIM unified flicker noise model.

Even more complex real-device models include the gate tunneling leakage model, the self-heating model, the floating body model, and the non-quasi-static model. The real-device models account for 80-90% of the model code, simulation time, and the model development effort. They are responsible for the accuracy of the compact model and the IC simulation. Many of the BSIM real-device models developed years ago are still valid for and are used in the FinFET standard compact model. These are described in Chapters 4–6.

Moving forward, the industry is trying to introduce germanium and InGaAs as the new channel material to enhance the carrier mobility. Will the BSIM model work with these advanced channel materials? BSIM researchers have found that with appropriate adjustments in model parameter values and small but important changes in model equations, the BSIM model works very well for the advanced channel material devices. For example, Figure 1.12 shows excellent BSIM model results for germanium FinFETs when a small improvement is made to the mobility model. The changes made are discussed in more detail in [15]. Similarly, excellent model results are obtained for InGaAs FinFETs as shown in Figure 1.13 [16].

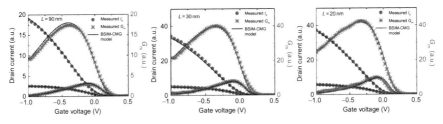

FIGURE 1.12

Modeling germanium FinFETs with the BSIM-CMG model for devices with channel lengths $L = 90$, 30, and 20 nm. With small but important improvements to real-device effects models, the BSIM-CMG model works well for advanced channel materials.

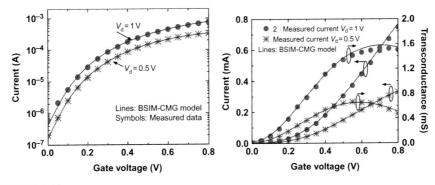

FIGURE 1.13

Transfer characteristics of an InGaAs FinFET modeled with the BSIM-CMG model shown on a semilog (left) and linear (right) scale showing the subthreshold and strong inversion region drain-current model. Experimental data are from Ref. [6].

1.8 THE INDUSTRY STANDARD FinFET COMPACT MODEL

Anticipating the adoption by industry of FinFET technology, the BSIM team started to develop a FinFET compact model in the mid-2000s, and released it in 2008. It is called BSIM-CMG. CMG stands for "common multigate." "Common" in "common multigate" means that all the multiple gates are electrically connected and share a common gate voltage. Multigate is a popular description for the FinFET in its various flavors—double gate, triple gate, quadruple gate. BSIM-CMG even models nanowire or pillar-shaped transistors with a cylindrical gate shape. These different FinFET flavors can be modeled by selecting the model switch GEOMOD appropriately as described in Chapter 3. There can also be a variation in the substrate on which FinFETs are manufactured. It can be a bulk or an SOI substrate. BSIM-CMG can model FinFETs on both of these substrates via the model select switch BULKMOD.

After the industry decided to embrace FinFET technology for production, BSIM-CMG was selected as the industry standard FinFET model, without contest, by the Compact Model Council. The Compact Model Council members are major IC companies, fabless companies, design automation companies, and IC foundries, including Altera, Analog Devices, Broadcom, Cadence, Globalfoundries, IBM, Intel, Mentor Graphics, Qualcomm, Renesas, Samsung, Hynix, ST Microelectronics, Synopsys, TSMC, and UMC. BSIM-CMG timely provided the foundation for the design infrastructure that enables FinFET-based circuit design and product development. The following chapters discuss all the important aspects of the standard FinFET model.

REFERENCES

[1] C. Hu, Modern Semiconductor Devices for Integrated Circuits, Pearson/Prentice Hall, New Jersey, 2010 (Chapter 7).

[2] Z. Liu, et al., Threshold voltage model for deep-submicrometer. MOSFET's, IEEE Trans. Electron Dev. 40 (1) (1993) 86–95.

[3] X. Huang, et al., Sub 50-nm FinFET: PMOS, IEDM Technical Digest, 1999, p. 67.

[4] A. Sachid, C. Hu, Denser and more stable FinFET SRAM using multiple fin heights, International Semiconductor Device Research Symposium (ISDRS), 2011, pp. 1–2.

[5] F.-L. Yang, et al., 5 nm-gate nanowire FinFET, VLSI Technology Symposium, 2004, pp. 196–197.

[6] J.J. Gu, X.W. Wang, H. Wu, J. Shao, A.T. Neal, M.J. Manfra, R.G. Gordon, P.D. Ye, 20-80 nm channel length InGaAs gate-all-around nanowire MOSFETs with EOT = 1.2 nm and lowest SS = 63 mV/dec, International Electron Devices Meeting, IEDM, 2012, pp. 27.6.1–27.6.4.

[7] B. Duriez, G. Vellianitis, M.J.H. van Dal, G. Doornbos, R. Oxland, K.K. Bhuwalka, M. Holland, Y.S. Chang, C.H. Hsieh, K.M. Yin, Y.C. See, M. Passlack, C.H. Diaz, Scaled p-channel Ge FinFET with optimized gate stack and record performance integrated on 300 mm silicon wafers, Electron Devices Meeting (IEDM), 2013, pp. 20.1.1–20.1.4.

[8] C.H. Wann, K. Noda, T. Tanaka, M. Yoshida, C. Hu, A comparative study of advanced MOSFET concepts, IEEE Trans. Electron Dev. 43 (10) (1996) 1742–1753.

[9] Y.-K. Choi, et al., Ultrathin-body SOI MOSFET for deep-sub-tenth micron era, IEEE Electron Dev. Lett. 21 (5) (2000) 254.

[10] B.J. Sheu, D.L. Scharfetter, C. Hu, D.O. Pederson, A compact IGFET charge model, IEEE Trans. Circuits Syst. 31 (8) (1984) 745–748.

[11] M.C. Jeng, P.K. Ko, C. Hu, A deep submicron MOSFET model for analog/digital circuit simulations, Technical Digest of International Electron Devices Meeting (IEDM), San Francisco, CA, 1988, pp. 114–117.

[12] J.H. Huang, Z.H. Liu, M.C. Jeng, K. Hui, M. Chan, P.K. Ko, C. Hu, BSIM3 Manual, University of California, Berkeley, 1993.

[13] T.Y. Chan, J. Chen, P.K. Ko, C. Hu, The impact of gate-induced drain leakage current on MOSFET scaling, Technical Digest of International Electron Devices Meeting (IEDM), Washington, D.C., 1987, pp. 718–721.

[14] K.K. Hung, P.K. Ko, C. Hu, Y.C. Cheng, A unified model for the flicker noise in metal-oxide-semiconductor field-effect transistors, IEEE Trans. Electron Dev. 37 (3) (1990) 654–665.

[15] S. Khandelwal, J.P. Duarte, Y.S. Chauhan, C. Hu, Modeling 20-nm germanium FinFET with the industry standard FinFET model, IEEE Electron Dev. Lett. 35 (7) (2014) 711–713.

[16] S. Khandelwal, J.P. Duarte, N. Paydavosi, Y.S. Chauhan, M. Si, J.J. Gu, P.D. Ye, C. Hu, InGaAs FinFET modeling with industry standard compact model BSIM-CMG, Nanotech 2014.

Compact models for analog and RF applications

2

CHAPTER OUTLINE

2.1 Introduction 15
2.2 Important compact model metrics 16
2.3 Analog metrics 16
    2.3.1 Quiescent operating point 17
    2.3.2 Geometric scalability 19
    2.3.3 Variability model.. 23
    2.3.4 Intrinsic voltage gain 24
    2.3.5 Speed: Unity gain frequency 31
    2.3.6 Noise 32
    2.3.7 Linearity and symmetry... 36
    2.3.8 Symmetry.. 42
2.4 RF metrics...44
    2.4.1 Two-port parameters 44
    2.4.2 The need for speed 46
    2.4.3 Non-quasi-static model.. 55
    2.4.4 Noise 58
    2.4.5 Linearity 64
2.5 Conclusion 68
References...68

2.1 INTRODUCTION

A compact model is the interface between the process technology and the circuit designer. The compact model, usually in the form of a design kit, encapsulates all the relevant details of the process in an easily accessible form that allows the designer to run simulations in a familiar environment, assessing the impact of the device on important circuit and system parameters, without having to understand the details of

FinFET Modeling for IC Simulation and Design. http://dx.doi.org/10.1016/B978-0-12-420031-9.00002-6

the physics or implementation details of the model. In fact, most designers have no idea what is inside the compact model, but they expect the model to capture the device behavior as accurately as possible. An analog designer will expect accurate prediction of bias points, small-signal parameters such as transconductance, output resistance, and capacitance, thermal and flicker noise, and device-matching properties. Since designers have no control over the design of a device, they can only vary the operating point and the device geometry, so they expect good physical scalability in the model. RF design requires accurate prediction of power gain, noise (thermal and flicker), linearity, and power.

FinFET devices come in an age where CMOS technology is playing a vital role in both analog and RF/microwave circuits. While technology scaling has improved the device speed by orders of magnitude, allowing operation in the millimeter-wave and subterahertz regime, voltage scaling and lower intrinsic gain has been a constant challenge to analog design. In such a world, most designers are averse to moving to advanced technology nodes since often the digital blocks are the only ones that gain in area and power. In this regard, a FinFET device can offer a change, improved output resistance (and hence intrinsic gain) and low capacitance. Given the high cost of doing fabrication runs in advanced technology nodes, an analog/RF compatible FinFET model will offer a unique opportunity to explore designs without incurring any costs of fabrication.

In this chapter, we will highlight important performance metrics for analog and RF circuits and relate these to the various compact model modules that are found in the FinFET compact model. This chapter therefore serves as a transition to the rest of the book, highlighting the importance of the various modules and the properties of a compact model.

2.2 IMPORTANT COMPACT MODEL METRICS

We begin by identifying important metrics that can be used to judge a compact model. Without these metrics, it would be difficult to gauge the accuracy of a compact model with regard to important quantifiable aspects that relate closely to circuit design. Metrics capture the essence of the model, and they should be important for the designer; in other words, they should be related to the performance of important circuit building blocks. The metrics are divided into analog and RF categories. It goes without saying that RF metrics are a superset and depend on analog metrics. In other words, an RF compact model should first and foremost satisfy the analog criteria for evaluation as they will form the core foundation for the model.

2.3 ANALOG METRICS

Classic analog design is concerned primarily with the gain-bandwidth product of amplifiers. This is due to the wide application of amplifiers (operational and transconductance amplifiers) as core building blocks and the use of feedback to linearize these blocks, making their performance predictable over the process and with temperature variation. Device currents are selected to minimize power while meeting speed requirements. Moreover, noise is of paramount importance in analog circuits as it determines the resolution, or lower end of the dynamic range, of the signal processing chain. The upper end of the dynamic range is determined by a voltage swing, which is related to the supply voltage and the linearity of the amplifier (and hence transistors). In many cases, linearity is a secondary concern because of the application of feedback, which inherently linearizes stages by trading gain for linearity. Owing to the lower supply voltage and lower intrinsic gain, the loop gain of a closed-loop system is dropping with technology scaling, making nonlinearity a more prominent concern even in analog applications. In the other extreme, people are exploring open loop design by relying on digital signal processing techniques to measure and correct for analog errors.

In fact, analog design is increasingly mixed-signal design. Pure analog design is very rare, or mostly confined to high-voltage applications. A FinFET device will live in a very mixed-signal environment, where sample time and charge processing will take place side by side with traditional linear voltage/current analog techniques. Moreover, FinFET devices will form the core of switches for the purpose of tuning and calibration, and often devices in these applications will experience voltage excursions that invert the device polarity (source-drain switch), requiring model symmetry. In sampled time systems, signal-dependent charge injection is important for quantifying the nonlinearity and systematic errors in analog signal processing.

2.3.1 QUIESCENT OPERATING POINT

In many respects, analog design comes down to biasing transistors correctly.[1] Circuit topologies are often selected on the basis of performance requirements, and the job of the circuit designer is to ensure that the transistors remain at the proper quiescent operation point with process variation and temperature variation. This is often a challenging task, as supply voltages scale to 1 V, and there is often very little margin, forcing designers to use biasing schemes that can compensate for variations in the device threshold voltage V_T with counter variations of another device. This means that they will trust the model's ability to track temperature or process variations (say, in the transistor geometric dimensions and doping). This places a high burden on a compact model since its current-voltage (I-V) behavior must be built on a physical foundation that inherently can predict these changes. Often students of compact modeling ask why one cannot simply reduce compact modeling to a mathematical exercise in curve fitting. The answer is partly that a mathematical model that fits the I-V curves from measured data without knowledge of any device physics would fail

---

[1] A saying attributed to Barrie Gilbert.

to predict the *I-V* curve for a small change in a physical parameter, such as doping or temperature. In theory, all these variations could be captured by a compact model, but technology foundries cannot afford to run thousands of combinatorial experiments to capture these variations, and instead rely on the compact model to reproduce these effects.

This reliance on physical trends is in contrast to the requirement of absolute accuracy. Absolute accuracy is fictional since the actual device of a sample will differ from the test structures used to measure the *I-V* curves. This process variation is an inevitable fact of fabricating integrated circuits with microscopic dimensions. In fact, it is remarkable that we are able to reproduce device behavior to the level of accuracy measured in the field, given that modern minimum-sized devices are limited by doping variation due to only a small number of dopant atoms. In fact, performance is more fundamentally becoming limited by quantum mechanical fluctuations due to the small dimensions and small number of charge carriers and dopants.

Most compact models are built from a physical core which is derived using a long-channel pseudo-two-dimensional transistor. The transistor charge *Q-V* characteristics are derived using only vertical fields, and currents are in turn derived by introducing a tangential field along the channel. The advantage of this approach is that it allows a closed-form equation to be derived without resorting to numerical integration. The most accurate solutions are the so-called surface potential solutions, which relate device charge to the surface potential along the channel. The downside is that the relationship between surface potential and device terminal voltages is not explicit, requiring Newton-Raphson style iteration to determine the solution. Moreover, the sensitivity to the surface potential is quite high, requiring very accurate convergence. But careful study of the convergence properties of the core equations allows these iterations to terminate in a fixed number of iterations, in effect unrolling the loops. The advantage of these "toy" models is that they are physically derived, and hence can capture the physical behavior of the device over the voltage excursions from accumulation, depletion, and weak inversion to moderate and strong inversion. Most importantly, the model can predict *I-V* and *Q-V* behavior in various stages of inversion, from weak to moderate inversion and to strong inversion. The moderate region is increasingly important for analog circuit operation since it allows the designer to trade off speed for gain and high transconductance efficiency (see below).

Another important aspect of the compact model is its ability to self-saturate. In other words, as the drain voltage is varied from tens of millivolts to the full supply voltage across the device, the saturation in the current should happen in a smooth and natural manner. In a classical long-channel transistor, saturation occurs owing to the so-called pinch-off effect. In a modern device, saturation occurs earlier owing to velocity saturation of carriers, arising from the high-field effects. Once a device is saturated, the *I-V* curve variation is due to the output conductance of the device, which is an accumulated sum of various complex mechanisms such as channel length modulation, drain-induced barrier lowering (DIBL), and impact ionization. Correct prediction of the curvature in this regime of operation is critical for analog circuits since the intrinsic gain will be determined by the slope of the I_{ds}-V_{ds} curve.

Finally, leakage currents play a critical role in discrete-time and low-power applications. Similarly to digital circuits, leakage currents are a concern when the device is "off," which happens when it is biased well below the threshold voltage. A real device will conduct very little current in this regime, but such a small current can quickly discharge a small capacitor, which introduces an error when sampling and storing a voltage on a capacitor.

How does one quantify these observations? The I-V curve of a given transistor with fixed dimension (W and L) reveals the most salient features. For example, a plot of I_{ds} versus V_{gs} for a family of V_{ds} (Figure 2.1a) quickly reveals how well a model can predict the device current with absolute bias voltage. If a device is biased in weak or moderate inversion, then the logarithmic plot is more useful as it expands this regime of operation (Figure 2.1b). A plot of the current versus drain-source voltage or a plot of I_{ds} versus V_{ds} for a family of V_{gs} (Figure 2.2) shows the current saturation behavior of a compact model and the accuracy of the model in saturation, which is critical for predicting the output conductance. The output conductance is more easily observed using small-signal parameters as discussed shortly, but certain trends can be observed even from the raw I-V curves, such as predictions of higher or lower current, which are consistently observed even from the I-V curves. Even though absolute accuracy is not important, such a consistent discrepancy reveals that certain device physics are missing in the compact model (or the device extraction was done incompletely).

If the device has a bulk terminal, then the body bias effect should also be included in these plots by varying the body bias and observing the predictability of the model. In this plot, we observe a shift in the threshold voltage of the device. If two independent gates are available, these curves should be plotted for both gates. In practice, many "back gates" will be used for leakage control (threshold shift) and may not be able to invert the channel, so the plots are very similar to those for bulk devices. For "common-gate" devices without a body contact, the plot of V_{gs} is then the only relevant curve. In these cases, it may be important to measure the I-V curves using pulsed mode measurements to avoid self-heating. Even though most devices have thin fins that are fully depleted, under the right conditions partial depletion may introduce kinks in the I-V curve that should be modeled correctly. DC and pulsed measurements with various back-gate/body biasing should reveal these device intricacies and the model should be verified to reproduce these effects.

2.3.2 GEOMETRIC SCALABILITY

Aside from the bias point, the only other variable under the control of a designer is the transistor geometry. Analog circuits make extensive use of L and W to trade off

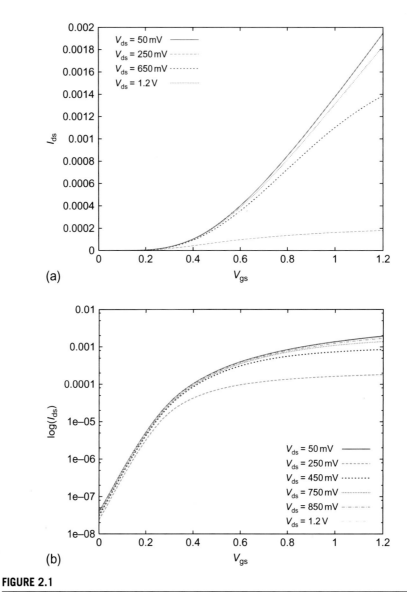

FIGURE 2.1

(a) I_{ds} versus V_{gs} for a family of V_{ds} and (b) log I_{ds} versus V_{gs} for a family of V_{ds}.

gain for speed, and unlike digital circuits, nonminimal values of L are used in many circuits. Moreover, it is not uncommon for the circuit designer to sweep W and L of key transistors to optimize performance. This means that a compact model must be extremely scalable with both W and L. In fact, a good designer would use a fixed W and sweep the number of transistor fingers N_f, which for FinFETs is equivalent

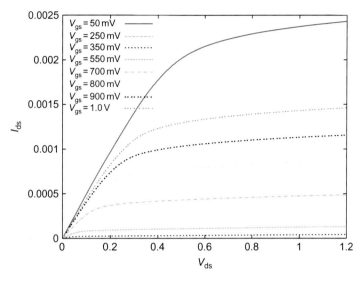

FIGURE 2.2

I_{ds} versus V_{ds} for a family of V_{gs}.

to scaling the number of fins. In this regard, a FinFET scaling is more constrained but naturally follows the preferred technique to scale a device (when matching is a concern).

In the extremes of scaling, short-channel effects are a well-known physical phenomenon that must be very well addressed by any nanoscale model. The complex two-dimensional (or three-dimensional) electrical field pattern along the channel and source/drain region impacts the device effective threshold voltage V_T, particularly because of nonuniform doping profiles. Scaling the length also gives rise to more pronounced DIBL, which is improved in the double gate structure of a FinFET. A plot of V_T versus the channel length (Figure 2.3a) is an important metric for a compact model, because the threshold voltage plays such a paramount role in circuit design. The importance of V_T is easily appreciated if one observes that the behavior of a FET is directly related to the amount of channel inversion (channel charge), which is directly impacted by V_T. This is why analog designers are concerned with the "overdrive" voltage (V_{gs}-V_T for a long-channel device), since it normalizes the gate bias relative to the threshold, which varies from device to device and as a function of temperature. Interestingly, in most FinFET processes, only a minimum-channel length is allowed, implying that longer-channel-length devices can be realized only by stacking devices. For all short-channel devices, the impact of DIBL is very important and can be understood from Figure 2.3b, the variation (reduction) in threshold voltage with increasing drain bias. One of the key advantages of the FinFET device is a reduction in DIBL owing to the strong-channel control afforded by two gates.

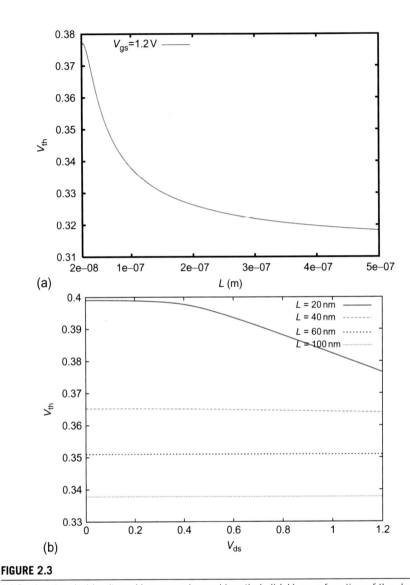

FIGURE 2.3

(a) Device threshold voltage V_T versus channel length L. (b) V_T as a function of the drain voltage V_{ds} for a family of channel lengths L.

Since transistors are three-dimensional structures, the results of geometric scaling are not trivial, especially when one considers the transistor extrinsic parasitic elements, such as device resistance and capacitance. In fact, a FinFET device has more capacitance than a comparable bulk device, and it is important to give the designer the maximum flexibility to scale the device to optimize the performance. Finally, the impact of well proximity is a well-known phenomenon in bulk devices, which forces designers to use many dummy fingers in their layouts to reduce the impact on the variation in threshold voltage and mobility. These effects, which are mostly extrinsic to the device, must be well modeled and captured. Given the smaller dimensions of a FinFET and its three-dimensional structure, the impact of stress and strain is even more pronounced, and can have both a positive and a detrimental impact on the device.

2.3.3 VARIABILITY MODEL

It is extremely important to capture device variation. Variation has many sources, including systematic and random fluctuations. Systematic variations include layout-dependent effects, such as the variation in the threshold voltage V_T as a function of the distance to the well, whereas random variations arise from natural variation in doping profiles and lithographic variations. These variations are usually categorized into device-to-device mismatch for closely spaced transistors, die-to-die variations, and wafer-to-wafer variations, often just called process variation. In other words, we are interested in both the variation for a single transistor from die to die and also that between a pair of "matched" transistors within the same die. Since circuit yield is a strong function of the variation of transistor performance, designers go to great lengths to ensure that their circuits operate over all possible process, voltage, and temperature variations. A common practice is to deliver transistor parameters for particular corners, such as "fast," "nominal," and "slow" corners for transistors (and other devices such as resistors and capacitors). If we just take transistor variations into account, since two flavors of transistors are used (n-type MOS and p-type MOS), this results in five separate corners that must be checked, often called SS, FF, SF, FS, and TT ("S" for "slow," "F" for "fast," and "T" for "typical").

While corner models are prevalent in industry, they are actually overpessimistic in most circumstances. The corner cases often represent worst-case conditions that occur very rarely in practice (Six Sigma), and designing circuits to meet corner cases requires overdesign, resulting in larger area and higher current consumption. A more attractive approach is statistical variation, where the model cards are derived by varying the actual physical parameters that vary, such as doping levels and lithographic dimensions. This underscores the need for compact models that are physically derived so that the statistical variation in basic physical parameters leads to correct variation in the model card parameters. When the model is nonphysical, it is very difficult to come up with a set of basic parameters to vary in the model card to match observed measurements of transistor variations. Even using principal component analysis on compact models with nonphysical origin in equations has not yielded satisfactory results.

An accurate mismatch model is another key model requirement. In fact, one may argue that mismatch is ultimately the biggest enemy of an analog circuit designer. The reason for this is due to the need for high levels of common-mode rejection and supply insensitivity in analog circuits. For example, the circuit in Figure 2.4 is a typical building block in amplifiers and filters in analog circuits. It is realized as a

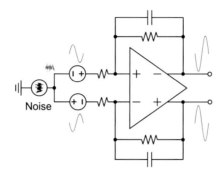

FIGURE 2.4

A fully differential circuit can reject supply noise and other common-mode noise if the circuit is perfectly balanced.

differential circuit in order to reject variations of the common mode of the signal and the power supply variations in the circuit. Power supply noise is inevitable, especially when circuits live in a mixed-signal environment where digital circuits create supply and substrate noise. If the circuit is perfectly balanced, this noise appears equally on the plus and minus sides of the amplifier, and can be rejected when the output signal is measured differentially, requiring good common-mode rejection. Any mismatch between the positive and negative side of the amplifier creates imbalance, limiting the amount of rejection. Moreover, mismatch in these differential circuits creates a DC offset that must be canceled, which introduces additional complexity in the circuit.

To fight mismatch, analog circuit designers scale transistor dimensions and bias points, or completely reject one technology node in favor of another. A good mismatch model again relies on a physically derived compact model that can accurately predict variations in key parameters such as V_T with variations in doping and lithography.

2.3.4 INTRINSIC VOLTAGE GAIN

One of the most important metrics in analog design is the DC gain of an amplifier, which is limited by the device output resistance in modern devices. As shown in Figure 2.5, a key building block such as a differential stage has a gain that is related to the product of the transconductance g_m and the output resistance r_o of the transistor. In reality, both the output resistance of the n-type MOS and the p-type MOS must be taken into account, so this metric represents a maximum upper bound on the gain of a single-stage amplifier:

$$A_0 = g_m(r_{on}||r_{op}) < g_m r_{on} = \frac{g_m}{g_{ds}}. \tag{2.1}$$

DC gain is important because it determines the accuracy of a closed-loop amplifier. One desires a high DC gain so that the loop gain of a multistage feedback amplifier is

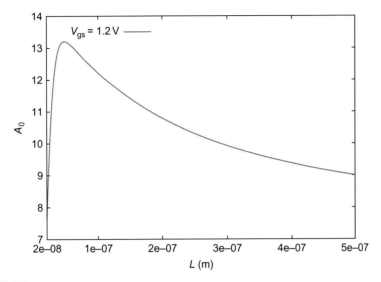

FIGURE 2.5

Device intrinsic gain A_0 plotted as a function of channel length L.

maximized. A high value of the loop gain translates into a large rejection of unwanted distortion generated by the amplifier. It also ensures high DC accuracy that is largely independent of temperature and the process.

DC gain is a good metric for compact model evaluation since it is the ratio of two small-signal parameters, which depend on the first derivatives of the current/voltage relations in the device. It is not a function of the device width (to first order) since as we scale W, both g_m and g_o are increased in the same proportion. It is a strong function of L, the channel length, owing to the impact on both g_m and g_o.

FinFETs offer a clear advantage over bulk devices because they offer higher intrinsic gain for the same channel length, mainly due to the reduction in DIBL as noted above, since a dual gate structure can maintain better channel control over the drain. This is why Equation (2.1) is recast as the ratio of g_m (gate control) and g_{ds} (drain control), which amplifies this point.

It is interesting to observe that the device intrinsic gain depends on the device bias current. Taking simple square law equations for simplicity, we have

$$A_0 = \frac{g_m}{g_{ds}} = \frac{2I_{ds}/V^*}{I_{ds}/V_A} = \frac{2V_A}{V^*}. \tag{2.2}$$

In the above equation, we have related the transconductance to the device overdrive voltage V^* and the "early voltage" V_A. For a long-channel device, V^* is equal to $V_{ds,sat}$, or the drain-source saturation voltage. This would be the value of the drain voltage which causes pinch off in the channel owing to a zero bias at the drain side $V_{ds,sat} = V_{gs} - V_T$. In a short-channel device, owing to high-field effects, $V_{ds,sat} \neq V^*$, so we can define V^* as the ratio of I_{ds} and g_m:

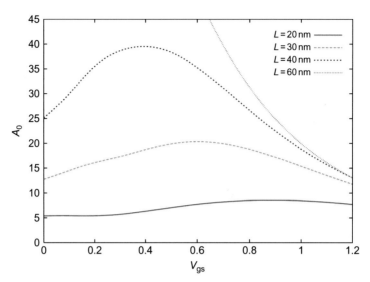

FIGURE 2.6

Device intrinsic gain A_0 versus V_{gs}.

$$V^* = \frac{2I_{ds}}{g_m}. \tag{2.3}$$

The important point is that the intrinsic gain is maximized by using the optimum overdrive voltage as shown in Figure 2.6. As the current in the device is reduced, the device eventually leaves the strong-inversion region, and as we approach weak inversion, the device drain current becomes an exponential functional of the gate-source bias, similar to a bipolar device, which implies that V^* approaches a constant related to the thermal voltage:

$$V^* = \frac{2I_{ds}}{(1/(nV_t))I_{ds}} = 2nV_t = 2n\frac{kT}{q}, \tag{2.4}$$

where $n < 1$ relates to how the gate voltage controls the surface potential through the voltage divider,

$$n = \frac{C_{ox}}{C_{ox} + C_{dep}}. \tag{2.5}$$

In the limit $n \to 1$, V^* approaches about 50 mV at room temperature. A plot of g_m/I_{ds}, sometimes known as the transconductance efficiency, versus V_{gs} is shown in Figure 2.7. As expected, the device bias current normalized transconductance peaks for low values of overdrive, or operation in weak inversion. Plotting the inverse relation (scaled by 2) gives us the effective overdrive voltage V^* (Figure 2.7b). For comparison, the square law relationship is also plotted, clearly showing a deviation as V_{gs} drops below the threshold voltage.

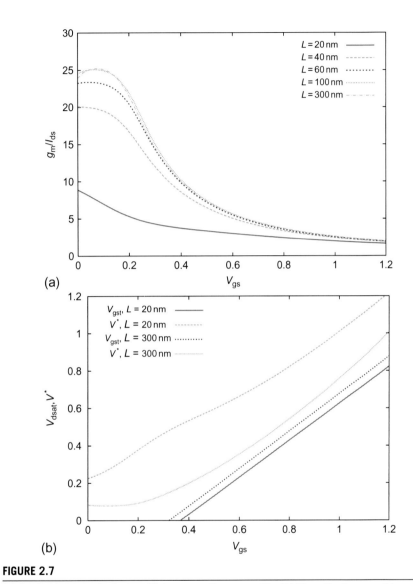

FIGURE 2.7

(a) Device transconductance efficiency g_m/I_{ds} versus V_{gs}. (b) The effective overdrive voltage V^* compared with V_{gs}-V_T. For the longer-channel device, the plots match fairly well, but for a short-channel device, they are very different.

The intrinsic gain is often plotted as a function of V_{ds} for a family of channel lengths L (Figure 2.8). For low values of V_{ds}, the device is in the triode region and the output resistance is low, resulting in low gain. Once V_{ds} causes saturation to occur, the output resistance increases and the gain maximizes. For large values of V_{ds},

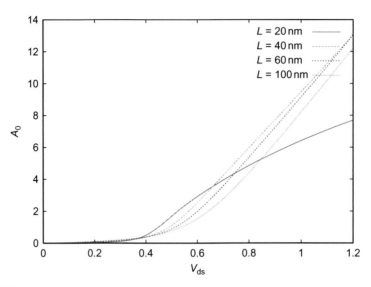

FIGURE 2.8

Device intrinsic gain A_0 versus V_{ds} for a family of channel lengths L.

detrimental effects of high drain fields such as DIBL and impact ionization currents once again reduce the gain.

To see these relations more explicitly, we can plot g_m as a function of V_{gs} and V_{ds} as shown in Figure 2.9. We observe that as the gate overdrive increases, the absolute value of g_m increases. For the diffusion component of current, at low overdrive, the increase in g_m is exponential owing to the exponential dependence of current on gate voltage. Operating with a higher bias increases the slope of the I_{ds}/V_{gs} curve in proportion to the increase in current. On the other hand, if we consider the drift component of the current, we know the current is proportional to the amount of charge in the channel and the carrier velocity. The channel charge scales in proportion to the gate bias, so for a fixed mobility, we expect g_m to increase linearly. In fact, for low fields, owing to Coulomb scattering, the mobility will experience an enhancement owing to screening provided by the inversion layer, and the increase in g_m is faster than linear. On the other hand, owing to high-field effects, we know the mobility will eventually degrade as carriers are pushed closer to the silicon-insulator interface, where surface scattering causes the mobility to drop. The trend for g_m as a function of V_{ds} can be divided into two regions: (1) in the triode region of operation, increasing V_{ds} increases the current, so the overall g_m increases; (2) as the device nears saturation, one would expect g_m to saturate.

In short-channel FETs, the output resistance $r_o = 1/g_{ds}$ has a strong bias dependence, especially on the drain voltage (Figure 2.10). The device output conductance in the triode region is very large since the drain voltage has a direct impact on the channel charge. In fact, to first order, the device g_{ds} is equal to the device g_m in this

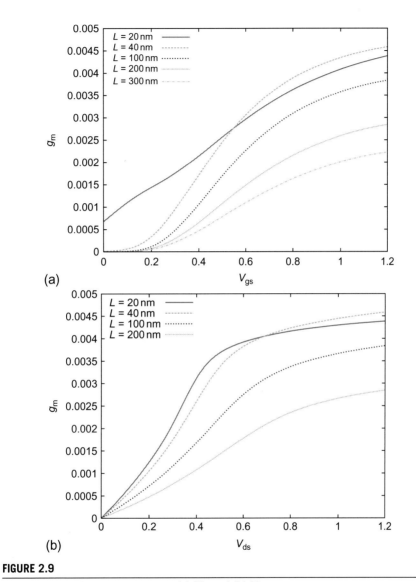

FIGURE 2.9

Device transconductance g_m versus (a) V_{gs} and (b) V_{ds}.

region, since varying the gate has the same impact as varying the gate owing to the presence of the inversion layer, making the drain terminal just as effective as the gate source. In a long-channel device, in the so-called pinch-off region, the device terminal no longer controls the channel charge directly, and the output conductance drops dramatically. In practice, the drain still modulates the depletion region width, which indirectly impacts the device through channel length modulation, which causes the

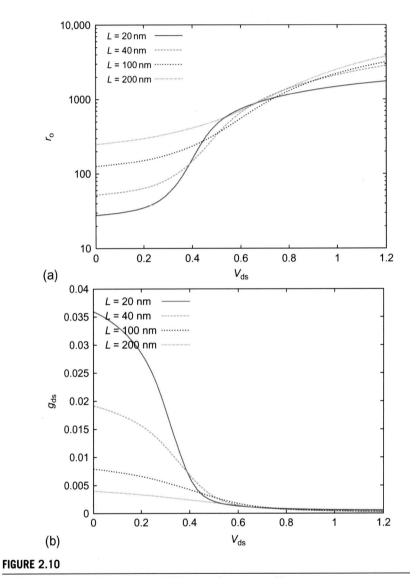

FIGURE 2.10

(a) Device output resistance r_0 and (b) $g_{ds} = 1/r_0$ versus V_{ds}.

current to increase with increasing drain bias, thereby increasing g_{ds}. The drain terminal has an even greater impact in short-channel devices, both because of the increase in the relative magnitude of the change in channel length δL relative to L, and because of DIBL. Owing to the complex two-dimensional field pattern in a short-channel device, a high voltage on the drain creates a high-field region that stores charge in the depletion region. This charge sharing in turn means that more electrons can flow into the channel, thereby lowering the threshold of the device. Fortunately, this effect is reduced for FinFETs, resulting in an improvement in intrinsic gain. Finally, as the voltage is further increased, impact ionization in the high-field region causes a further increase in current and thereby a drop in the output resistance.

2.3.5 SPEED: UNITY GAIN FREQUENCY

In analog circuits, DC gain is only half the story. While operation at subthreshold is beneficial for gain and especially transconductance efficiency, in practice this is rarely acceptable because of the reduction in the device speed. Only extremely slow circuits can tolerate operating in this region. For most other circuits, operation in moderate or strong inversion is desired.

The speed of a transistor is often measured using the device unity gain frequency f_T, or the frequency when the current gain of a transistor is unity. Since $i_d = g_m v_{gs}$, and $v_{gs} j\omega (C_{gs} + C_{gd}) = i_s$, taking the ratio, we have

$$A_i = \frac{i_d}{i_s} = \frac{g_m}{j\omega(C_{gs} + C_{gd})}. \tag{2.6}$$

Solving for $|A_i| = 1$, we arrive at the unity gain frequency

$$\omega_T = 2\pi f_T = \frac{g_m}{C_{gs} + C_{gd}}. \tag{2.7}$$

It is not at first obvious why f_T plays such an important role in analog (and digital) circuits. To see this, consider the simple cascade amplifier shown in Figure 2.11a, where a single-stage amplifier drives an identical copy. The gain of this first stage is given by the intrinsic gain of the amplifier, and the 3 dB bandwidth is limited by the pole at the high impedance node:

$$\omega_0 = \frac{1}{r_{o,1}((1 + |A_2|)C_{gs,2} + C_{d1,tot})}. \tag{2.8}$$

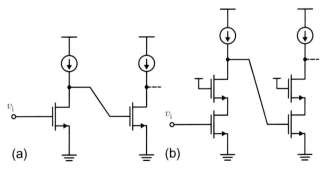

(a)   (b)

FIGURE 2.11

(a) A cascade of common-source amplifiers. (b) A cascade of cascode amplifiers.

Here $C_{d1,tot}$ is the total drain capacitance ($C_{gd} + C_{ds} + C_{wire} + \cdots$), and the effect of Miller multiplication is captured by boosting the second-stage input capacitance by the gain A_2. If the second stage is a cascode with a small voltage gain between the gate and the drain, then A_2 can be made smaller than unity to minimize its impact. So if we disregard the Miller effect, we have

$$\omega_0 = \frac{1}{r_o(C_{gs} + C_{d,tot})},\tag{2.9}$$

as an upper bound. In most applications, we intend to embed this amplifier in a feedback loop, so the gain-bandwidth product of interest is

$$A_0\omega_0 < g_m r_o \frac{1}{r_o(C_{gs} + C_{d,tot})} = \frac{g_m}{C_{gs} + C_{d,tot}} \approx \omega_T,\tag{2.10}$$

which bears a resemblance to the unity gain of the transistor itself. So in an approximate fashion, we can see that the maximum gain-bandwidth product of an amplifier is in fact limited by the device unity gain frequency. Even in a digital circuit, we find that f_T plays a key role. Consider the time constant of a simple gate, such as an inverter. If the fan out of an inverter is unity, in other words the inverter drives an identical copy of itself, then the load of the inverter is approximately $C_{gs} + C_{d,tot}$. The discharge current takes a complicated form, but to first order the transistor acts like a switch with on-conductance $g_{ds} = g_m$ (in the triode region), so the discharge time is given by

$$\tau_T = \frac{C_{gs}}{g_m} = \frac{1}{\omega_T}.\tag{2.11}$$

A plot of f_T as a function of V_{gs} for different channel lengths is shown in Figure 2.12. The bias and geometry dependency on g_m have already been discussed. The remaining dependency is on the device capacitance, including the gate-to-source capacitance C_{gs} and the gate-to-drain capacitance. It is important to note that f_T is usually measured using scattering parameters, and so using the "DC" values of g_m and C will fail to capture f_T at high frequency. We will return to this point in Section 2.4.

In a long-channel device, C_{gs} is usually the dominant capacitor, since C_{gd} occurs only from the small overlap between the gate terminal and the drain region. In a real device, a considerable portion of the capacitance may be due to extrinsic capacitors in the wiring. In modern devices, the gate-drain overlap region is a significant fraction and C_{gd} can be comparable to C_{gs}. In a FinFET, owing to the three-dimensional structure, a considerable amount of fringing capacitance increases C_{gd} compared with a bulk device. For these reasons, FinFET f_T suffers compared with that of an equivalent bulk device. In reality, bulk devices cannot scale to FinFET dimensions, so the comparison is not fair. Nevertheless, this requires a very good extrinsic FET model to accompany the intrinsic model.

2.3.6 NOISE

Thermal noise is relatively well understood in FETs and should form the core of any model. When drift current dominates, the thermal noise is a function of the channel

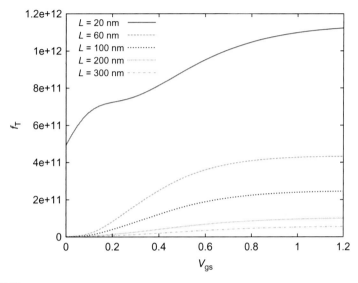

FIGURE 2.12

The device unity gain frequency f_T as a function of V_{gs} for a family of channel lengths L.

conductance, whereas in moderate and weak inversion the diffusion component gives rise to shot noise. A good model should capture the drain noise current accurately in all regions in a smooth and uniform manner. Capturing noise in saturation is complicated by the excess noise generated by the FET, captured by the parameter γ in the spot noise variance of the drain current:

$$\overline{i_{d,n}^2} = 4kTg_{ds,ch}\gamma\delta f, \tag{2.12}$$

where $g_{ds,ch}$ is the channel conductance. For a long-channel device, it is well known that $\gamma = 2/3$, which is derived by considering the thermal noise contribution of an incremental region of the channel to the drain. The coupling to the gate through the gate-oxide capacitance also gives rise to a correlated gate noise current. For short-channel devices it has been observed that γ is larger than 1, which is sometimes attributed to the hot electron effects occurring on the drain end of the transistor. A good test for a compact model is a plot of γ versus the drain and gate bias voltage, as shown in Figure 2.13. It is important to define γ in terms of $g_{ds,ch}$ (the drain-source conductance in the "linear region" of the device) and not g_m, which leads to the erroneous conclusion that γ is much larger than 1. The reason is that noise arises from the physical resistance of the channel, which originates from the inversion layer at the source to a region near the drain. As noted before, $g_m = g_{ds}$ in the linear region of the transistor operation.

One would expect that the thermal noise of a FET varies smoothly as the device bias is varied from strong inversion to weak inversion, with the limits of weak-inversion noise given by the well-known shot noise current $2qI_{ds}$. For strong-inversion

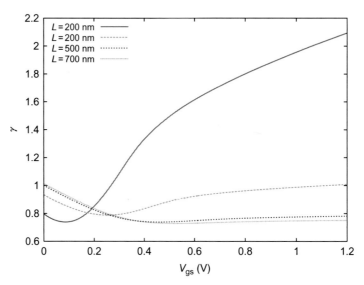

FIGURE 2.13

Excess noise factor γ versus gate bias V_{gs}.

operation, a plot of γ versus V_{ds} should have physical meaning in the saturation regime. To understand the importance of γ, let us derive the noise figure of a common-source and a common-gate amplifier. For the common-source case at low frequencies, the total output noise is given by

$$\overline{i_o^2} = g_m^2 \overline{v_{R_s}^2} + \overline{i_d^2} = g_m^2 4kT \, R_s \delta f + 4kT \, g_{ds,ch} \gamma \delta f, \tag{2.13}$$

from which we calculate the noise figure

$$F = 1 + \frac{g_{ds,ch} \gamma}{g_m^2 R_s} = 1 + \frac{\gamma/\alpha}{g_m R_s}. \tag{2.14}$$

For many applications, R_s is fixed, and g_m is limited by power consumption, which shows the noise limit is ultimately determined by γ. The common-gate amplifier is even more sensitive to γ since it has no current gain,

$$\overline{i_o^2} = \overline{i_{R_s}^2} + \overline{i_d^2} = 4kTG_s\delta f + 4kTg_{ds,ch}\gamma\delta f, \tag{2.15}$$

giving the well-known lower bound on the obtainable noise figure

$$F = 1 + \frac{(g_m/\alpha)\gamma}{G_s}. \tag{2.16}$$

The reason the common-gate amplifier is favored over the common-source amplifier is the ability to obtain a wideband impedance match by setting $g_m = R_s$, which results in

$$F = 1 + \frac{\gamma}{\alpha}. \tag{2.17}$$

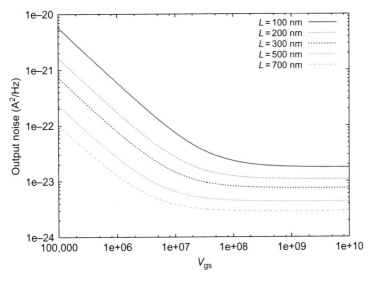

FIGURE 2.14

The drain noise current spectral density versus frequency for a family of channel lengths L biased at the same bias V_{gs}.

Later we shall also see that the ultimate lower limit on phase noise is similarly related to γ.

In addition to thermal noise, flicker noise plays a big role in analog circuits, especially as devices are devices are scaled to smaller dimensions. The spectrum of flicker noise in a FET overlaps with the frequency of interest in many analog signal processing applications (DC to 10 s or MHz), and a designer will often size devices in order to minimize the impact of flicker noise (Figure 2.14). The flicker noise model should therefore be very physical in scaling limits, particularly when the designer selects different channel lengths L to minimize noise. Naturally, the bias dependence of flicker noise should be correctly captured.

In a FET, all of the intrinsic noise originates from the channel and traps in the gate-oxide interface. Extrinsic parasitic elements, such as physical resistance in the gate, source, and drain, also contribute noise. Plotting noise parameters versus bias is a good way to tease out the various sources of noise: ones that depend on bias are generally due to intrinsic device noise, and fixed noise contributions are due to extrinsic sources. An alternative representation of the noise of a FET is to define an input pair of correlated noise voltage and current sources. This is useful for circuit design and also a good metric for testing the model. Some circuit simulators can simulate and plot these parameters for a two-port circuit, and these plots are a good way to test the physical variation in the input noise sources, which depend on both intrinsic and extrinsic noise sources.

2.3.7 LINEARITY AND SYMMETRY

Harmonic distortion

Linearity, or the ability of an analog circuit to faithfully reproduce the input signal at the output (amplified or buffered), is an important metric, although it receives less attention than gain and speed. The reason is simple, since most analog circuits employ feedback, which has the ability to reject the distortion produced by devices by the loop gain of the circuit. Open loop amplifiers are much commoner in RF circuits, so intrinsic device nonlinearity is still an important metric. Even in modern nanoscale devices, linearity is playing a bigger role as loop gains drop (owing to the drop in intrinsic gain). Furthermore, when FETs are used as switches, particularly in sampled time systems, the distortion produced by the device is sampled and processed along with the signal.

The source of distortion is easy to understand since the *I-V* characteristics of a FET are described by nonlinear equations of terminal voltages:

$$I_{ds} = f(V_{gs}, V_{ds}, V_{bs}). \tag{2.18}$$

Expanding the above relation into a Taylor series of V_{gs}, while holding V_{ds} and V_{bs} constant, and disregarding the DC, we have the following "G_m" nonlinearity power series:

$$i_{ds} = g_{m1}v_{gs} + g_{m2}v_{gs}^2 + g_{m3}v_{gs}^3 + \cdots, \tag{2.19}$$

where lowercase symbols denote AC quantities. The first term, g_{m1}, is nothing but the transconductance of the device. The curvature of the *I-V* curve gives rise to higher-order coefficients which describe the deviation of AC from the quiescent operating point for a small deviation of the gate-source voltage about the bias point. In a similar way, we can derive a power series in terms of the drain-source voltage,

$$i_{ds} = g_{ds1}v_{ds} + g_{ds2}v_{ds}^2 + g_{ds3}v_{ds}^3 + \cdots, \tag{2.20}$$

and, in general, a two-dimensional power series can be derived when both the gate-source voltage and the drain-source voltage vary. In practice, to keep things simple, it is common to deal only with one-dimensional power series relations.

The commonest way to measure the nonlinearity of a block is to find the harmonic distortion (HD), which is defined by driving an amplifier with a pure sinusoid and observing the harmonics at the output:

$$i_{ds} = g_{m1}A_{gs}\cos(\omega t) + g_{m2}A_{gs}^2\cos^2(\omega t) + g_{m3}^3 A_{gs}^3 \cos^3(\omega t) + \cdots. \tag{2.21}$$

For example, $\cos^2(x) = 0.5(1 + \cos(2x))$, or the square term produces DC and the second harmonic. Likewise, cubing the signal gives

$$\cos^3(x) = \cos(x)\cos^2(x) = 0.5(\cos(x) + \cos(x)\cos(2x)) \tag{2.22}$$

$$= \frac{1}{2}\left(\cos(x) + \frac{1}{2}(\cos(x) + \cos(3x))\right) \tag{2.23}$$

$$= \frac{3}{4}\cos(x) + \frac{1}{4}\cos(3x), \tag{2.24}$$

which contains the fundamental signal and the third harmonic.

In general, with use of a binomial expansion on the sum of complex exponentials, it is not very difficult to show that the nth odd power produces harmonics starting from the fundamental and every odd harmonic (first, third, fifth, ..., nth), whereas an nth even power produces all even harmonics starting from DC (zero, second, fourth, sixth, ..., nth). In other words, even harmonics generate DC and every even harmonic up to n ($0 \times \omega, 2 \times \omega, \ldots, n \times \omega$). Odd harmonics generate the fundamental and every odd harmonic up to n ($1 \times \omega, 3 \times \omega, \ldots, n \times \omega$). This allows us to write the following equation for the harmonics of the output drain current when the input is excited by a sinusoid:

$$i_{ds} = \left(g_{m1} A_{gs} + \frac{3}{4} g_{m3} A_{gs}^3 + \cdots \right) \cos(\omega t) \tag{2.25a}$$

$$+ \left(\frac{1}{2} g_{m2} A_{gs}^2 + \cdots \right) \cos(2\omega t) \tag{2.25b}$$

$$+ \left(\frac{1}{4} g_{m3} A_{gs}^3 + \cdots \right) \cos(3\omega t) + \cdots . \tag{2.25c}$$

The important point is that each harmonic arises from a potentially infinite set of powers, but for small excursions from the bias point, the most dominant terms are the smallest powers. This last point is important, and may seem like a contraction, but in fact it follows from the assumption of a "small-signal" or "weak" distortion. For example, for a practical device, when the equation is put into a power series, one finds that the relevant quantity is a normalized voltage v_{gs}/V^* (overdrive voltage), or in weak inversion, $v_{gs}q/kT$, so as long as the normalized voltage excursion is small, raising this to an increasingly larger power produces an increasingly smaller quantity in the power series. Under these conditions, the lowest-order powers completely dominate the behavior of the device distortion generation.

Given these conditions for weak distortion, the HD is defined as the ratio of the amplitude of the harmonic to the fundamental. For example,

$$HD_2 = \frac{V_{2\omega}}{V_\omega} = \frac{g_{m2}(A_{gs}^2/2)}{g_{m1} A_{gs}} = \frac{1}{2} \frac{g_{m2}}{g_{m1}} A_{gs} \tag{2.26}$$

and

$$HD_3 = \frac{V_{3\omega}}{V_\omega} = \frac{g_{m3}(A_{gs}^3/4)}{g_{m1} A_{gs}} = \frac{1}{4} \frac{g_{m3}}{g_{m1}} A_{gs}^2. \tag{2.27}$$

As shown in Figure 2.15, as we sweep the amplitude of the input sinusoid, the magnitude of the second harmonic increases quadratically, whereas the third harmonic grows cubically, so their HD ratios exhibit a power law with one order less (linear and quadratic). As the amplitude grows beyond the small-signal distortion regime, eventually higher-order terms dominate and the slopes change.

To show the importance of overdrive voltage, plots of HD_2 and HD_3 versus V_{gs} are shown in Figure 2.16. The physical reason for the improvement in HD with V_{gs} is quite simply related to the fact that the electric fields in the device are related to the overdrive voltage, and lower fields produce more linear relations as expected. The dip

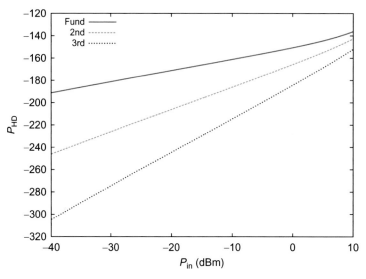

FIGURE 2.15

The HD currents at 2ω and 3ω versus the input amplitude P_{in} (dBm) applied at the gate/source showing the characteristic slope of 2 and 3 on a log scale. Note the bias is fixed at V_{gs}.

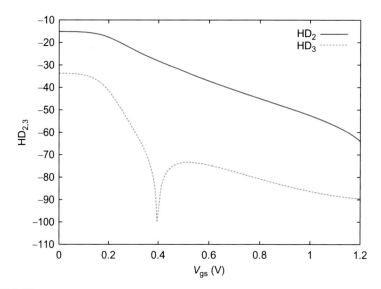

FIGURE 2.16

HD_2 and HD_3 versus the bias V_{gs} for a fixed swing v_{gs} (-30 dBm). Note the "sweet spot" for HD_3.

in HD$_3$ is a consequence of $g_{m3} = 0$ for a particular bias point, sometimes called the "sweet spot" of a MOSFET.

While we have defined HD for G_m nonlinearity, the same metrics can be defined for any other pair of currents/voltages in the device. While G_m nonlinearity is the commonest type of nonlinearity, output conductance nonlinearity is perhaps a close contender. Since HD is a strong function of the device bias point, it is hard to define a universal metric. The modeling engineer should ensure that the model reproduces the distortion at a range of different bias voltages (both V_{gs} and V_{ds}), and most importantly, it is important to ensure that the slope of the HD products follows the above relations for weak inputs. For example, on a log scale, the third harmonic signal should have a slope of 3 when the input is very weak, and any deviation from this slope indicates a problem in the compact model.

This is because we expect all the compact model curves to be smooth and infinitely differentiable, which means that we can always define a power series and prove that the slope is indeed 3. But if the compact model has a source of an abrupt nonlinearity, then a nonsmooth curve will not follow our predictions. A classic example is an "if" statement in the compact model that changes the device I-V behavior (e.g., a diode model that has a hard saturation). If the signal is biased near this "if" statement, then even a small deviation produces a kink in the I-V transfer (a point where the slope is not defined). Keep in mind that for a strong enough signal, this assumption breaks down since the fundamental power is affected by all odd harmonics, which causes the slope to change dramatically, but this occurs only when a device is driven very hard (deep compression).

While it is not possible to check every possible HD product, it is good practice to check HD$_2$ through HD$_5$ to make sure all curves follow the correct slopes predicted by small-signal distortion. Since distortion arises from the large signal behavior of a transistor, it has to be simulated using transient or steady-state simulation, such as periodic steady-state analysis using shooting in SpectreRF or harmonic balance simulation. The same can be obtained from a transient simulation by calculating the fast Fourier transform of the output waveform and plotting harmonic powers. Classic SPICE also includes a disto statement that is similar to an AC analysis, but includes the effect of higher-order harmonics as we have done.

Gain compression

A very important metric is the gain compression that naturally occurs in all devices owing to nonlinearity. For example, if we plot the effective transconductance G_m of a device versus the input swing, we obtain a curve as shown in Figure 2.17. As expected, for small input voltages v_{gs}, $G_m = g_m$, but as we increase the swing, the transconductance either increases or drops. The reason for this change in G_m is related to the fact that all odd harmonics of the I-V relation generate fundamental current ($g_{m,2k-1}$ for $k > 1$ in our power series), and these currents can be in phase or out of phase with the fundamental generated by g_{m1}. If we examine Equation (2.25), we find that the fundamental output current amplitude is given by

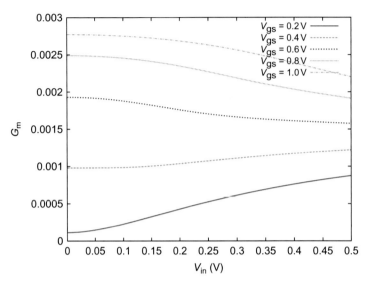

FIGURE 2.17

The transconductance gain compression versus the signal swing v_{gs}.

$$i_{ds,\omega} = g_{m1} v_{gs} + \frac{3}{4} g_{m3} v_{gs}^3 + \cdots .$$ (2.28)

If we normalize this current by the gate drive amplitude v_{gs}, we obtain the effective transconductance

$$G_m = \frac{i_{ds,\omega}}{v_{gs}} = g_{m1} + \frac{3}{4} g_{m3} v_{gs}^2 + \cdots .$$ (2.29)

So we see that the effective transconductance is related to the gate-source voltage excursion v_{gs}. For very small swings, in fact the small-signal limit for $v_{gs} \to 0$, then $G_m = g_{m1}$ as expected. But for larger swings, depending on the sign of g_{m3}, the transconductance can increase or decrease. The increase in transconductance, or more generally gain, is called gain expansion. Likewise, a decrease in gain is called gain compression. In the plot in Figure 2.17, when the device is biased in weak inversion, the gain is expansive owing to the exponential nature of the device *I-V* curve. For higher bias points, the gain is compressive owing to the high-field mobility effects in the FET.

Note that we have explicitly separated the gain into its components G_m and R_o since gain expansion also occurs owing to the limited output swing (headroom) and output resistance R_o of a device. In practice, both act in conjunction to determine the overall gain compression characteristics.

Since all transistors operate with a fixed headroom, eventually the voltage swing must be clipped by the rails, and then the gain therefore drops, leading to an eventual compressive characteristic (Figure 2.18). But the shape of the compression, in

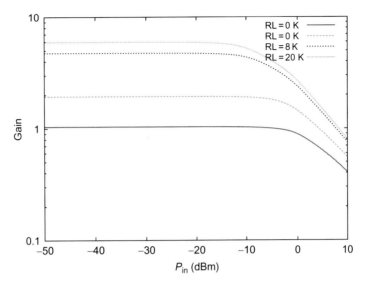

FIGURE 2.18

The voltage gain compression versus the signal swing P_{in} (dBm). The drain is biased at 1.2 V using an ideal inductor and the load resistor R_L is varied. Note that high values of R_L have higher gain, but compress earlier owing to the drain voltage swing (rather than the G_m compression).

particular regions of expansion and compression, is related to the device nonlinearity and is a good way to gauge the accuracy of a compact model. For example, G_m compression occurs due to high-field effects and mobility degradation. While these effects are visible on the I-V curve as well, here the effects are measured by the slopes of the I-V curve, which may not be as apparent to the naked eye.

Memory effects

To keep things simple, we have explicitly disregarded all capacitive nonlinearity in the FET. This means that we have assumed that the drain current is an instantaneous function of the gate-source and drain-source voltages, which is not true in practice. Owing to the presence of both linear and nonlinear capacitance at the terminals of a FET, the situation is much more complicated, requiring a Volterra series description instead of a power series relation [1]. In practice, in most analog circuits, these so-called memory effects are disregarded. This is justified in practice if the device is operated well below the cutoff frequency of the transistor, allowing the quasi-static assumption to be valid, and well below the poles of the amplifier, so that the amplitudes of signals are not attenuated or signals are not phase shifted as the signals travel through a transistor. These conditions are often met for the desired signals that are processed by analog circuits, but not for the harmonics.

Intermodulation distortion

In practice, it is rare to amplify a single tone, and in general a complex signal is processed by a device. For linear time-invariant circuits, we know that we need only to characterize the frequency response of the system (or equivalently the impulse response in the time domain), but for nonlinear circuits, even memoryless ones, we have to characterize the impact of not only every tone, but every two tones, three tones, and in general N tones, to completely understand the response of the system to a general signal. Owing to the complexity of doing this, a commoner metric is to characterize the intermodulation between only a few tones. The commonest metrics are the two-tone response and the associated intermodulation signals (in general not harmonically related to either tone). We will return to intermodulation distortion in Section 2.4.

2.3.8 SYMMETRY

The symmetry of a compact model is also an important metric in analog circuits that use the device bias around $V_{ds} = 0\,V$, which occurs when the device is used as a switch in sampling circuits and in commutating circuits (such as RF mixers). The origin of the asymmetry in the compact model is not the drain-source swapping in the model, as was once believed, nor is it due to source referencing, but asymmetry can arise from both of these mechanisms if they are not handled properly. For example, often an "if" statement is used to swap the internal source/drain of a physically symmetric FET, since the label "source" and "drain" are defined only by the bias points ($V_{ds} > 0\,V$ for an n-type FET), which may change sign during the course of a circuit simulation. The model typically assigns one terminal as the drain and detects a change in sign, which can be done only with a finite precision. In other words, when $V_{ds} < \epsilon$, for some small ϵ, the drain-source swapping will occur. This will produce a small kink in the transfer curve about $V_{ds} = 0\,V$, which requires smoothing.

The Gummel symmetry test is used to detect potential symmetry problems in a FET model. The circuit shown in Figure 2.19 is used to excite the circuit about

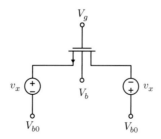

FIGURE 2.19

The circuit used to characterize the Gummel symmetry test.

$V_{ds} = 0$ V. For a uniformly doped FET, we expect the drain current to be an odd function of v_x,

$$I_d(v_x) = -I_d(-v_x), \tag{2.30}$$

so

$$i_x(v_x) = \frac{1}{2}(I_d(v_x) - I_s(v_x)). \tag{2.31}$$

For a properly designed compact model, we expect i_x is a smooth function of v_x with continuity at $v_x = 0$ V, implying that $d^n i_x / dv_x^n$ for all n. From physical considerations, we expect $i_x = 0$ A at $v_x = 0$ V, and also that $d^2 i_x / dv_x^2 = 0$ due to the symmetry of waveform. For example, first- and second-order derivatives for the FinFET common multigate (CMG) model are plotted as shown in Figure 2.20, clearly demonstrating smooth and consistent behavior.

In addition to the current behavior, the terminal charge behavior of the Gummel symmetry circuit should also satisfy physical conditions on symmetry and smoothness. Sweeping v_x from a negative value to an equal positive value, one can plot the gate, source, drain, and bulk charges of the MOSFET (Q_g, Q_s, Q_d, and Q_b), respectively. The derivatives of charges versus v_x should be continuous around $v_x = 0$ V, and the derivatives of charges should pass smoothly through $v_x = 0$ V and have no discontinuities. Furthermore, the derivatives of Q_s and Q_d should be equal and opposite at $v_x = 0$ V because the device is assumed to be symmetric.

An ideal symmetric device should also exhibit symmetry in the capacitances C_{gs} and C_{gd}, with equal values at $v_x = 0$ V. Similar comments apply to C_{bs} and C_{bd}, and C_{dd} and C_{ss}, respectively.

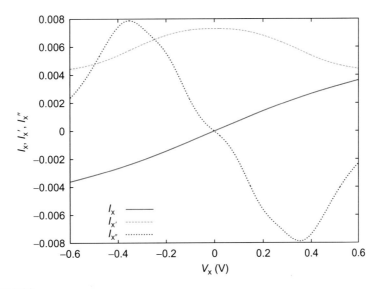

FIGURE 2.20

The Gummel symmetry test for the BSIM-CMG FinFET model.

2.4 RF METRICS

A good analog compact model is the foundation for a good RF compact model. In addition to the already mentioned important metrics for analog circuits, we find that RF circuits require new modules in the compact model to capture gate resistance, substrate resistance, and non-quasi-static (NQS) effects. Since many RF circuits do not use feedback, they are more sensitive to distortion produced by the transistors, placing stronger demands on the accuracy of the model. Finally, since inductors can be used to tune out capacitive parasitic elements, there are also lower tolerances for error in the capacitance of the device. Furthermore, we find that f_T is not the best metric for speed, although it continues to play an important role.

2.4.1 TWO-PORT PARAMETERS

At higher frequencies, it is much commoner to rely on two-port parameters of a transistor. The reasons for the importance of two-port parameters, as opposed to say a hybrid-pi model or a full compact model, are both historical and practical. In the past, compact models were relatively simple and did not capture the detailed device behavior and extrinsic parasitic elements, particularly for operation near f_T of the device when NQS effects are important to model. An RF model that includes the intrinsic device and packaging parasitic elements can become extremely complex for analysis by hand (Figure 2.21). To circumvent this issue, designers often relied on measured two-port parameters at high frequency, rather than the extrapolated model behavior from *I-V/C-V* curves, in order to understand the device behavior.

Two-port parameters have many limitations, most importantly arising from their small-signal assumption and their frequency dependence. But many RF circuits

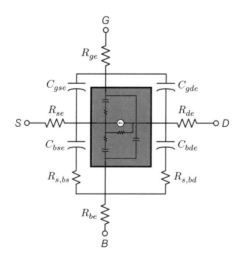

FIGURE 2.21

A full FET model includes the intrinsic transistor in addition to external parasitic elements.

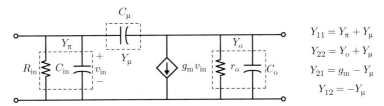

$$Y_{11} = Y_\pi + Y_\mu$$
$$Y_{22} = Y_o + Y_\mu$$
$$Y_{21} = g_m - Y_\mu$$
$$Y_{12} = -Y_\mu$$

FIGURE 2.22

The measured Y parameters can be converted directly into circuit pi parameters using the relations shown. Since this is an oversimplified model, the parameters are frequency dependent.

(power amplifiers are a good exception to this) operate with weak signals (signals captured off the air) over a narrow range of frequencies (so-called narrowband communication) and the small-signal narrowband assumption is valid in many circumstances, especially if there are no interfering signals. Moreover, by using measured two-port (or three/four-port) parameters over a range of bias points, on can construct a table-lookup model to capture the dynamics for modulating signals. The benefit of two-port parameters is that they are completely general and capture all of the transistor dynamics (by definition) at a particular frequency of interest. Moreover, they can be used to answer some very general questions about a transistor device, such as "What is the maximum possible power gain for a device?" For these reasons, we will make extensive use of two-port parameters and the rich theory to delve into these questions.

Two-port parameters from a measured device are often compared to a compact model as a metric, but often many details are hard to decipher directly from these parameters as both circuit designers and device engineers have more intuition for circuits and small-signal equivalent circuits. A good solution is to convert general two-port parameters into an equivalent circuit representation, such as the popular hybrid-pi model, in order to gain intuition about how well the model matches measurements. Keep in mind that the two-port parameters vary with frequency, so the equivalent circuit will also vary, but if the two-port is chosen wisely, then many of the parameters remain fairly constant. For example, if we consider the two-port parameters of a transistor in the common-source configuration, as shown in Figure 2.22, then it is easy to show that the equivalent pi circuit is related to the Y parameters as follows:

$$Y_{11} = Y_\pi + Y_\mu, \tag{2.32}$$

$$Y_{22} = Y_o + Y_\mu, \tag{2.33}$$

$$Y_{21} = g_m - Y_\mu, \tag{2.34}$$

$$Y_{12} = -Y_\mu. \tag{2.35}$$

We can solve these equations for four pi-circuit parameters from measured results:

$$Y_\pi = Y_{11} + Y_{12}, \tag{2.36}$$

$$Y_\mu = -Y_{12}, \tag{2.37}$$

$$Y_o = Y_{22} + Y_{12}, \tag{2.38}$$

$$G_m = Y_{21} - Y_{12}. \tag{2.39}$$

It is important to note that Y_μ and G_m are in general complex. The real part of C_μ reflects that the feedback in a transistor can experience a phase shift at high frequencies, and the imaginary part of G_m reflects the same in the forward direction. It is convenient to think of G_m in magnitude/phase terms,

$$G_m = |G_m| \angle G_m, \tag{2.40}$$

and it is also convenient to decompose the input network Y_π from a shunt representation to a series one, since in practice the gate resistance is fairly constant with frequency:

$$Y_\pi = j\omega C_\pi + G_\pi = \frac{1}{(1/(j\omega C_{gs})) + R_g}. \tag{2.41}$$

The output network, on the other hand, is already in a convenient form:

$$Y_o = j\omega C_{ds} + g_o. \tag{2.42}$$

Since these capacitors and resistors represent the entire transistor, they are in fact lumped representations of many sources of capacitance arising from both the intrinsic device and the extrinsic parasitic elements, including the substrate network. Note that this common-source representation cannot capture the model accuracy for source side parasitic elements, which may be different owing to layout differences. For this reason, a two-port representation is not complete and a three-port or four-port characterization is preferred (for the back gate or second gate of the device). For a typical device, a plot of the pi parameters is shown in Figure 2.23. Over a narrow range of frequencies, these parameters are relatively constant and a good indication of the device behavior for a fixed operating point.

2.4.2 THE NEED FOR SPEED

RF circuits work from low gigahertz to several thousand gigahertz, from the so-called RF spectrum, to the millimeter-wave spectrum, and even at "terahertz" frequencies. There is a quest to run circuits as fast as possible to exploit new bandwidth and to maximize communication rates. For analog circuits, we found that f_T was a good substitute for estimating the gain-bandwidth product of a circuit. But in RF circuits, capacitive parasitic elements can be tuned out and the transistor can be operated over a narrow bandwidth at frequencies in excess of f_T. So what is the right metric for a transistor?

Let us start with a simple exercise and derive the gain and bandwidth for the cascade of tuned amplifiers shown in Figure 2.24. Here the voltage gain per stage is limited not by the output resistance, but by the inductor quality factor Q_L:

$$A_v = g_m r_o || R_L \approx g_m Q_L \omega L. \tag{2.43}$$

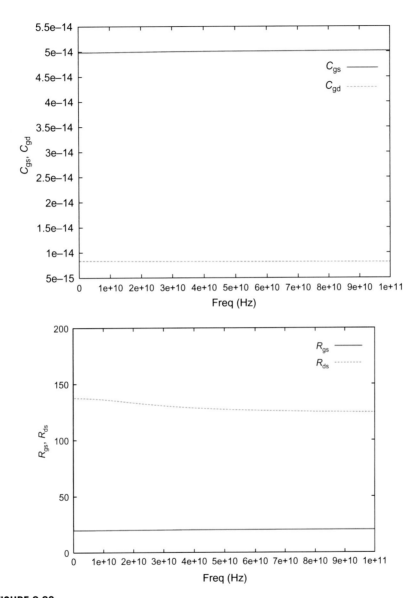

FIGURE 2.23

The capacitance and resistance pi parameters of a core device versus frequency. Over a wide range of frequency, these parameters are constant and correspond closely to the oft-used hybrid-pi equivalent circuit model.

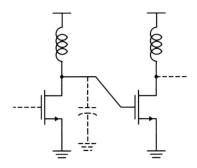

FIGURE 2.24

The cascade of tuned amplifier stages. l is chosen to tune out the capacitive parasitic elements at the drain node of each transistor at the desired operating frequency.

For a fixed device transconductance g_m, we wish to maximize the product of Q_L and the inductance, L. We wish to use the smallest possible inductors in order to minimize their physical area. The value of L is not arbitrary, though, and at minimum it must tune out the next-stage capacitance. So for simplicity, let us assume no explicit capacitance is added to the circuit so that $C_L = C_{gs} + C_{gd}$. To minimize C_{gd}, we can use a cascode stage as shown. So to first order,

$$A_v = g_m Q_L \omega L = g_m Q_L \frac{1}{\omega C_{gs}(1 + \mu)} = \frac{g_m}{C_{gs}(1 + \mu)} \frac{Q_L}{\omega}. \tag{2.44}$$

Interestingly, we see that f_T has found its way into our equation,

$$A_v \approx \frac{f_T}{f} Q_L, \tag{2.45}$$

which states that our center frequency f times the gain is limited by the f_T and Q_L product. This drives the well-known point of the importance role played by the passive devices at RF frequencies. But is this equation fundamental? No, since we just took a fixed topology common-source amplifier and found the voltage gain per stage in a cascade of identical stages.

A more fundamental question is given a transistor, what is the maximum gain that one can obtain under the optimal conditions? A related question is, what the maximum frequency where one can in fact get power gain from a transistor?

A unique aspect of RF circuits is that power gain is often more important than voltage gain, since RF circuits must drive (or are driven by) fixed impedances, often with a low value of $50\,\Omega$. In this case, a more relevant question is what is the most gain we can get out of a transistor? The well-known result from two-port theory is that the gain is maximized by biconjugate matching, or in other words by designing the source and load admittance to be conjugately matched with the input and output admittance of the two-port:

$$Y_s = Y_{in}^* = \left(y_{11} - \frac{y_{12} z_{21}}{y_{22} + Y_L} \right)^*, \tag{2.46}$$

$$Y_L = Y_{out}^* = \left(y_{22} - \frac{y_{12}y_{21}}{y_{11} + Y_s} \right)^*, \tag{2.47}$$

where y_{ij} are the two-port Y parameters of the two-port (transistor), and Y_s and Y_L are the source and load admittance. If these values are fixed (e.g., by the antenna port), then an impedance transformation network can be used to obtain the optimal Y_s and Y_L given above. A very interesting result is that if the above source/load impedance exists, then the maximum gain obtainable is given by

$$G_{max} = \frac{Y_{21}}{Y_{12}} \left(K - \sqrt{K^2 - 1} \right), \tag{2.48}$$

where $K \geq 1$ is the stability factor of the two-port:

$$K = \frac{2\Re(Y_{11})\Re(Y_{22}) - \Re(Y_{12}Y_{21})}{|Y_{12}Y_{21}|}. \tag{2.49}$$

A stable two-port has $K \geq 1$, which means that it will not oscillate for any physical load or source admittance ($\Re(Y_{S|L}) > 0$). By definition, an unstable two-port has infinite gain since it generates an output signal for a zero input, which is why we restrict our choice to $K \geq 1$. Since the gain increases as we reduce the stability factor, a common metric is the maximum stable gain (MSG):

$$MSG = G_{max}|_{K=1} = \frac{Y_{21}}{Y_{12}}. \tag{2.50}$$

The way to understand the MSG metric is to imagine that you have a two-port that is conditionally stable, so that $K < 1$. Instability implies that the input and/or output admittance of the two-port has a negative real part, $\Re(Y_{S|L}) < 0$. We can make this two-port stable by intentionally adding loss to its ports until the net real part is zero, $\Re(Y_{S|L}) = 0$. At this point, the two-port is conditionally stable and its gain is given by MSG. While adding any more loss makes the two-port stabler, it also reduces the gain.

An example may shed some light on this. Consider a FET transistor modeled by the simple hybrid-pi model shown in Figure 2.25. At very low frequency, the FET is unconditionally stable since C_μ provides negligible feedback. But for higher frequencies, for an inductive load, it is easy to see that the input admittance has a negative real part since the current flowing through the feedback capacitor has a quadrature phase relation with the voltage at the drain:

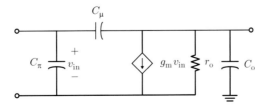

FIGURE 2.25

The hybrid-pi model for a simple FET transistor amplifier.

$$i_\mu = (v_g - v_d)j\omega C_\mu \approx (v_g - v_g(-g_m j\omega L))j\omega C_\mu \tag{2.51}$$

$$= v_g(j\omega C_\mu - g_m \omega^2 L C_\mu), \tag{2.52}$$

$$\Re(Y_{in}) = \Re\left(\frac{i_\pi + i_\mu}{v_g}\right) = -g_m \omega^2 L C_\mu. \tag{2.53}$$

A more exact relation can be derived directly from the complete two-port parameters:

$$Y_{in} = Y_{11} - \frac{Y_{12} Y_{21}}{j\omega L + Y_{22}}. \tag{2.54}$$

These relations can be simplified to

$$Y_{in} = j\omega(C_\pi + C_\mu) - \frac{-j\omega C_\mu g_m}{g_o + j\omega C_o + (1/(j\omega L))} \tag{2.55}$$

$$\approx j\omega(C_\pi + C_\mu) - \frac{-j\omega C_\mu g_m}{1/(j\omega L)}, \tag{2.56}$$

$$\Re(Y_{in}) \approx -g_m \omega^2 L C_\mu, \tag{2.57}$$

which matches our intuitive calculation above. The key point is that the approximation is valid in the frequency range when the inductive load presents a higher susceptance to the device output conductance and susceptance. This can easily happen in a real device. To make the device stable, one can either add gate resistance to the device or add shunt resistance (or series) in a manner to increase the real component of the load voltage compared with the imaginary part. By doing this, we can improve the stability factor to the point $K = 1$.

The maximum gain of a device is then dependent on the value of K. For $K < 1$, the maximum gain is obtained by stabilizing the device and is given by the MSG. For $K > 1$, the maximum gain is given by Equation (2.48). A plot of the maximum gain for a typical device, then, follows the curve shown in Figure 2.26. There is a kink in the curve at the point where K crosses unity as described.

For the example device, we can derive K from Equation (2.49);

$$K = \frac{\Re(j\omega C_\mu(g_m - j\omega C_\mu))}{\omega C_\mu \sqrt{g_m^2 + \omega^2 C_\mu^2}} \tag{2.58}$$

$$= \frac{\omega C_\mu}{\sqrt{g_m^2 + \omega^2 C_\mu^2}}. \tag{2.59}$$

For frequencies of interest, $\omega C_\mu \ll g_m$,

$$K \approx \frac{\omega C_\mu}{g_m} < 1, \tag{2.60}$$

which shows that the transistor is conditionally stable as predicted before. Note that we have disregarded the physical gate resistance in a real device, which changes the above calculation since we assumed $\Re(Y_{11}) = 0$.

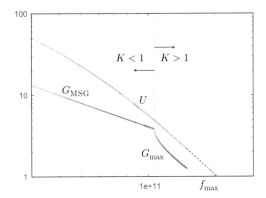

FIGURE 2.26

The maximum gain available from a transistor two-port has two regions: when $K < 1$, it is given by MSG, and for higher frequencies when $K > 1$, it is given by G_{max} or the available gain.

The maximum unity power gain frequency (f_{max})

The commonest figure of merit for a high-frequency performance of a transistor is the maximum frequency of oscillation, f_{max}, or equivalently, the highest frequency at which we can obtain power gain from a transistor. Let us work with a simple hybrid-pi model as before, but introduce gate resistance and disregard C_μ (for simplicity), as shown in Figure 2.27. In this case, owing to the unilateral nature ($Y_{12} = 0$), we can simply find the maximum gain by impedance matching the source and load directly:

$$Z_s = Z_{in}^* = \left(R_g + \frac{1}{j\omega C_{gs}} \right)^* = R_g + j\omega L_s, \tag{2.61}$$

$$Y_L = Y_{out}^* = (G_o + j\omega C_{ds})^* = G_o + j\omega L_d. \tag{2.62}$$

Under these matched conditions, the power gain is easily calculated without recourse to the full two-port parameters since the reactances are absorbed by the source/load as shown in Figure 2.27:

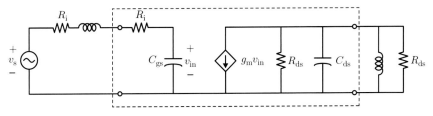

FIGURE 2.27

A simple model for a FET used to calculate the power gain. The model does not include feedback capacitance but includes gate resistance.

$$P_{\text{in}} = \frac{1}{2} V_{\text{in}} I_{\text{in}} = \frac{1}{2} \cdot \frac{v_s}{2} \cdot \frac{v_s}{2R_i} = \frac{1}{8} \frac{v_s^2}{R_i}, \tag{2.63}$$

$$P_L = \frac{v_o^2}{2r_o} = \frac{1}{2r_o} \left(\frac{g_m r_o}{2} \cdot v_\pi \right)^2 = \frac{1}{2r_o} \left(\frac{g_m r_o}{2} \cdot \frac{v_s}{2R_i} \frac{1}{\omega C_\pi} \right)^2. \tag{2.64}$$

Taking the ratio, and noting that $g_m / C_\pi \approx \omega_T$, we have a simple expression for the power gain:

$$P_L = \left(\frac{\omega_T}{\omega} \right)^2 \frac{r_o}{R_i^2} \frac{1}{32} v_s^2, \tag{2.65}$$

$$G_p = \frac{P_L}{P_{\text{in}}} = \frac{1}{4} \left(\frac{r_o}{R_i} \right) \left(\frac{\omega_T}{\omega} \right)^2. \tag{2.66}$$

Now we are in a position to find the maximum frequency f_{max} for which the transistor provides power gain:

$$G_p = 1 = \frac{1}{4} \left(\frac{r_o}{R_i} \right) \left(\frac{f_T}{f_{\text{max}}} \right)^2, \tag{2.67}$$

$$f_{\text{max}} = \frac{f_T}{2} \sqrt{\frac{r_o}{R_i}}. \tag{2.68}$$

Despite the simplicity of our transistor model, this equation is very insightful as it connects the maximum power gain to the device f_T, or unity gain frequency. This affirms our faith in f_T as an important metric for RF and analog circuits, but it also shows that f_T is not the complete story. The device f_{max} may in fact be larger than f_T if the output resistance r_o is a large factor bigger than the device input resistance R_i. For example, it can be shown that R_i depends on the gate polysilicon resistance and the channel resistance as seen from the gate terminal (gate AC travels through the gate oxide and flows out through the channel and into the source and drain terminals). In the limit, if we lay out the device with many fingers to minimize the physical gate resistance (see Figure 2.28), only the channel resistance contributes to R_i [2] (see Section 2.4.3 for more details):

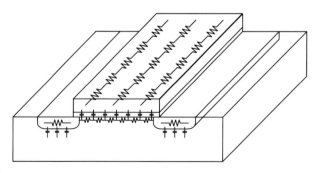

FIGURE 2.28

The origin of gate resistance has two components: from the physical gate material, which is minimized with a layout using multifinger devices, and the actual channel resistance, which is unavoidable and layout invariant.

$$R_i \approx \frac{1}{5g_m}, \tag{2.69}$$

so

$$f_{max} = \frac{f_T}{2}\sqrt{\frac{r_o}{1/(5g_m)}} = \frac{f_T}{2}\sqrt{5g_m r_o} = \frac{f_T}{2}\sqrt{5A_0}. \tag{2.70}$$

Since for any respectable transistor $A_0 > 1$, we have $f_{max} > f_T$. We have disregarded the feedback capacitor C_μ, which also plays a critical role. A more thorough analysis yields the following expression for f_{max} [3], which is very similar to the one adopted by the International Technology Roadmap for Semiconductors [4]:

$$f_{max} \approx \frac{f_T}{2\sqrt{R_g\left(g_m C_{gd}/C_{gg}\right) + \left(R_g + r_{ch} + R_s\right)g_{ds}}}, \tag{2.71}$$

which highlights the importance of minimizing the loss in the device, such as the drain resistance R_d, source resistance R_s, and gate resistance R_g (part of R_i in our hybrid-pi model, the rest coming from r_{ch}). These parasitic elements are in large part determined by the layout and the process technology, which is a good metric for testing both the intrinsic and the extrinsic transistor model.

In RF circuits we can neutralize this capacitance by connecting an inductor L_μ in parallel to resonate it out at any given frequency and over a narrow bandwidth. In differential circuits, we can neutralize the capacitance with a cross-coupled matching capacitor as shown in Figure 2.29. In fact, is the optimal L_μ the value that completely cancels the capacitance?

For a FET we know that $Y_{21} \approx g_m$ and $Y_{12} \approx j\omega C_{gd}$, so

$$\text{MSG} = \left|\frac{Y_{21}}{Y_{12}}\right| \approx \frac{g_m}{\omega C_{gd}}. \tag{2.72}$$

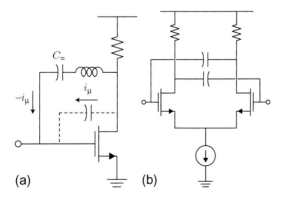

(a)   (b)

FIGURE 2.29

Two techniques to neutralize the feedback capacitance (C_{gd}) in a FET include adding (a) an inductor in shunt to resonate out the capacitance, and (b) alternatively in a fully differential or balanced amplifier, cross-coupling to produce a negative capacitance in shunt which works over a broad frequency range.

This implies that as we reduce the effective C_{gd} by neutralization, the gain increases:

$$\text{MSG} \approx \frac{g_m}{\omega(C_{gd} - C_n)}, \tag{2.73}$$

where C_n is the amount of neutralization. It can be shown that under these conditions

$$\text{MSG} = \sqrt{\frac{g_m^2}{\omega^2(C_{gd} - C_n)} + 1}, \tag{2.74}$$

with a corresponding stability factor given by

$$K = \left(1 + \frac{2g_g g_{ds}}{\omega^2(C_{gd} - C_n)^2}\right) \text{MSG}^{-1}. \tag{2.75}$$

Stability requires $K > 1$, which occurs for a range of neutralization values n:

$$n_1 \leq n \leq n_2, \tag{2.76}$$

where n is defined as the relative amount of neutralization

$$n = \frac{C_n}{C_{gd}}, \tag{2.77}$$

and the critical values of n occur at

$$n_1 = 1 - \frac{1}{\omega C_{gd}}\sqrt{\frac{g_g g_{ds}}{U - 1}}, \tag{2.78}$$

$$n_2 = 1 + \frac{1}{\omega C_{gd}}\sqrt{\frac{g_g g_{ds}}{U - 1}}, \tag{2.79}$$

where U is Mason's unilateral gain for the two-port, which will be discussed next. In fact, the maximum gain occurs at these critical points, rather than at $n = 1$, and the maximum gain can be computed to be [5]

$$G_{max} = 2U - 1. \tag{2.80}$$

So what is U and how does it relate to the power gain of the two-port?

Mason's unilateral gain U

One of the most important metrics in high-frequency transistor characterization is U, or Mason's unilateral gain [6]. The idea behind U has an interesting history, recounted in [7]. The definition of U is the power gain for a two-port under the general four-port lossless embedding shown in Figure 2.30. The idea is to provide lossless "feedback" to unilaterize the two-port, and then under these conditions U is the gain we obtain from the two-port:

$$U = \frac{|Y_{21} - Y_{12}|^2}{4(\Re(Y_{11})\Re(Y_{22}) - \Re(Y_{12})\Re(Y_{21}))}. \tag{2.81}$$

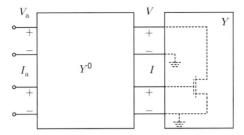

FIGURE 2.30

A two-port network Y is embedded into a four-port network Y^0 providing lossless feedback to unilaterize the device.

The U function has several important properties:

1. If $U > 1$, the two-port is active; if $U \leq 1$, the two-port is passive.
2. U is the maximum unilateral power gain of a device under a lossless reciprocal embedding.
3. U is the maximum gain of a three-terminal device regardless of the common terminal.

These properties have contributed to the widespread use of U as a metric to test a transistor power gain, and it is a good metric to test a compact model for the same reasons. Unlike MSG, U is extremely sensitive to the loss of a device. This is because even if $K < 1$, we can add a network around the device to make it unilateral, and hence stable, and then we can extract the MSG from the device in this configuration. We find that the less loss there is in a device, the more gain we can extract, in stark contrast to MSG, which in fact requires us to *add* loss to stabilize the device when $K < 1$. For these reasons, we find it much easier to match a model to data for MSG rather than U. U is a function of what really matters to obtain power gain, and it is therefore the metric of choice.

Is U the maximum possible power gain we can obtain? As we have already seen in the case of neutralization, it is not in fact the maximum. While neutralization provides a gain of $2U - 1$, it can be shown that the maximum possible gain is as high as [8]

$$G_{\text{max}} = 2U - 1 + 2\sqrt{U(U - 1)} \approx 4U. \tag{2.82}$$

2.4.3 NON-QUASI-STATIC MODEL

In the derivation of most compact models, the *I-V* relations and *C-V* relations are derived independently, with the implicit assumption that charges respond instantaneously to external voltages. This assumption is known as the quasi-static assumption, which states that if the voltages at the terminals of a device are varied slowly enough, then we would expect the *C-V* relations and *I-V* relations to accurately describe the current flow into the transistor. But how fast can we go and still consider this assumption valid? How long does it take charge to flow into the channel? A very naive calculation would say that the RC time constant associated with charging the channel is approximately the resistance of the channel r_{ds} times the gate-oxide capacitance C_{gs}. As we have seen, the channel resistance r_{ds} in the triode region, or equivalently the channel conductance g_{ds} in the triode region, is equal to g_m of the device in saturation. So with this approximation, the time constant is

$$\tau = \frac{1}{g_m} \times C_{gs} = \frac{C_{gs}}{g_m} = \frac{1}{\omega_T},$$ (2.83)

which shows that the device f_T is again playing an important role. It seems that if we approach the device f_T, our quasi-static assumption begins to fail.

To see this in a compact model, we can plot the real and imaginary parts of Y_{11}, where port 1 is the gate and port 2 is the drain. For a real device we would expect to see a real component of Y_{11} owing to the channel conductance and the distributed nature of the gate-to-channel capacitance. In a quasi-static model, though, the gate-source impedance is purely reactive and the real part of Y_{11} can arise only from feedback from C_{gd}. In Figure 2.31a, we demonstrate this by plotting $\Re(Y_{11})$ and the component R_{gs} of the pi model for a model with the NQS effect turned on and off (NQSMOD = 0, see Appendix A for description of parameters). It is clear that the NQS resistor has an important impact on the overall device. As we have seen, loss in the device plays an important role in determination of the maximum available gain from the device. Without this resistor, we have the nonphysical result that a device has infinite available gain (disregarding feedback), since the input would dissipate no power in a reactive load, whereas the output would generate power.

We can also see the frequency dependence of the NQS effect by observing the effective transconductance G_m versus frequency. We define the effective transconductance from the Y parameters as the forward minus the reverse short-circuit transconductance:

$$G_m = |Y_{21} - Y_{12}|.$$ (2.84)

For a physical device, we would expect that G_m should drop with frequency, but a simulation of a model without the NQS effect, shown in Figure 2.32, shows the opposite trend, predicting that the device transconductance gets better at high frequency. With the NQS model, the correct behavior is observed, which can be verified with measurements. Note the full segmented model described in [9] is needed to capture this effect fully.

Various techniques have been developed to model NQS effect with a SPICE model. Since SPICE is inherently an "ordinary differential equation" solver, it cannot intrinsically handle distributed components. To model distributed effects, we need to approximate the distributed nature using lumped circuits. For example, a relaxation time approximation models the charge deficit or surplus in the channel [10], which decays to its nominal quasi-static value with a time constant similar to $1/\omega_T$. The NQS model can also be lumped into a gate resistance model (and added to the

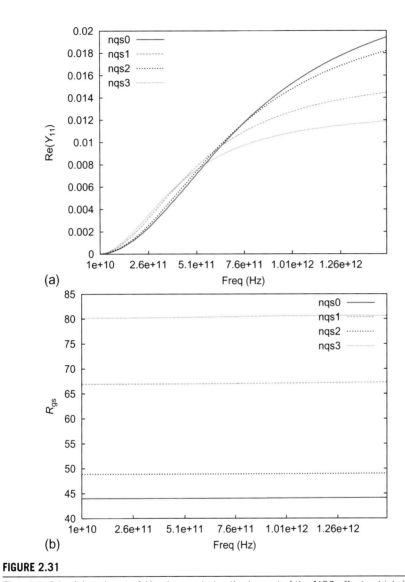

FIGURE 2.31

The plot of the (a) real part of Y_{11} demonstrates the impact of the NQS effect, which (b) is evident as an extra component of gate resistance R_{gs}.

physical gate resistance) [11]. The most accurate models use charge segmentation and the continuity relation to model the nonuniform charge distribution in the channel [12, 13]. These various models are compared in Figure 2.32, showing that each model has a frequency range over which it performs quite well. Only when devices are pushed beyond f_T is it necessary to invoke the NQS model.

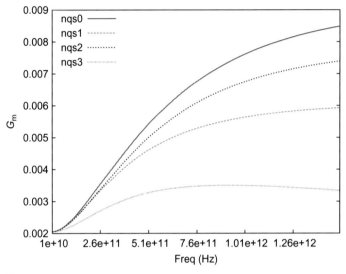

FIGURE 2.32

A comparison of various NQS models built into the BSIM-CMG model and their impact on the effective G_{m}.

2.4.4 NOISE

The noise performance of a two-port can be analyzed in general in terms of the equivalent input noise voltage/current sources and their correlation (Figure 2.33). Partitioning the input noise current into two components, a component correlated ("parallel") to the noise voltage and a component uncorrelated ("perpendicular") to the noise voltage, we have

$$i_n = i_c + i_u, \tag{2.85}$$

where we assume that $\langle i_u, v_n \rangle = 0$ and

$$i_c = Y_c v_n. \tag{2.86}$$

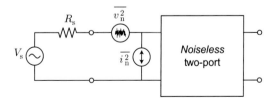

FIGURE 2.33

A general two-port noisy network can be partitioned into a noiseless two-port and two correlated noise sources and the inputs and/or the outputs. An input-referred two-port is shown here.

We can therefore write

$$v_{eq} = v_n(1 + Y_c Z_s) + Z_s i_u. \tag{2.87}$$

Minimum achievable noise figure (F_{min})

With this model, we can calculate the noise figure and minimize it by finding the optimal source impedance. Let $\overline{v_n^2} = 4kTBR_n$, $\overline{i_u^2} = 4kTBG_u$, and $\overline{v_s^2} = 4kTBR_s$. Then

$$F = 1 + \frac{R_n|1 + Y_c Z_s|^2 + |Z_s|^2 G_u}{R_s}. \tag{2.88}$$

If we let $Y_c = G_c + jB_c$, and $Y_s = Z_s^{-1} = G_s + jB_s$, it is not to difficult to show that the optimum source impedance to minimize F is given by

$$B_{opt} = B_s = -B_c, \tag{2.89}$$

$$G_{opt} = G_s = \sqrt{\frac{G_u}{R_n} + G_c^2}. \tag{2.90}$$

The minimum achievable noise figure is

$$F_{min} = 1 + 2G_c R_n + 2\sqrt{R_n G_u + G_c^2 R_n^2}. \tag{2.91}$$

Through some algebraic manipulations, one can derive [14]

$$F = F_{min} + R_n R_s \left| G_{opt} - G_s \right|^2. \tag{2.92}$$

R_n is called the noise sensitivity parameter. This terminology is clear since the rate of deviation from the optimal noise figure is determined by R_n. If a two-port has a small value of R_n, then we can be sloppy and sacrifice the noise match for gain. If R_n is large, though, we have to pay careful attention to the noise match. A plot of NF_{min} (logarithm of F_{min}) versus the device bias V_{gs} is shown in Figure 2.34. These plots were for a fixed frequency of 20 GHz and a family of channel lengths L are used to clearly show the trend that increasing the device f_T (minimize L) improves the achievable noise figure.

Simple model for FET noise

Consider the following noise sources associated with the source R_s, the gate resistance R_g, the channel resistance R_{ch}, and the load R_L:

$$\overline{v_s^2} = 4kT\,BR_s,$$

$$\overline{v_g^2} = 4kT\,BR_g,$$

$$\overline{i_d^2} = 4kT\,Bg_{d0}\gamma B,$$

$$\overline{i_L^2} = 4kT\,BG_L.$$

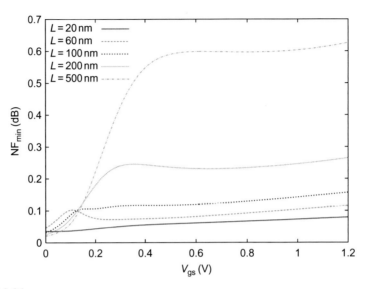

FIGURE 2.34

A plot of the minimum achievable FET noise figure versus bias (V_{gs}) for a family of channel lengths (L).

The gate noise can be used to model the channel noise seen on the gate side and the correlation is disregarded (in addition to the physical thermal noise on the gate side). This is known as the Pospiesalski noise model, and it matches measurements pretty well [15].

$$F = 1 + \frac{\overline{v_g^2}}{\overline{v_s^2}} + \frac{\overline{i_d^2} + \overline{i_L^2}}{g_m^2 \overline{v_s^2}} \tag{2.93}$$

$$= 1 + \frac{R_g}{R_s} + \frac{g_{d0}\gamma + G_L}{R_s g_m^2}. \tag{2.94}$$

If we assume $g_m = g_{d0}$ (long channel),

$$F = 1 + \frac{R_g}{R_s} + \frac{\gamma}{g_m R_s} + \frac{G_L G_s}{g_m^2}. \tag{2.95}$$

If we make g_m sufficiently large, the gate resistance will dominate the noise. The gate resistance has two components, the physical gate resistance and the induced channel resistance:

$$R_g = R_{poly} + \delta R_{ch} = \frac{1}{3}\frac{W}{L}R_\square + \frac{1}{5}\frac{1}{g_m}. \tag{2.96}$$

For the optimal source impedance, one can show that

$$F_{min} = 1 + 2\left(\frac{\omega}{\omega_T}\right)\sqrt{g_m R_g \gamma}. \tag{2.97}$$

Doing the best layout possible, we obtain the limit that $R_g = 1/5g_m$ and

$$F_{min} = 1 + 2 \left(\frac{\omega}{\omega_T} \right) \sqrt{\gamma/5}. \qquad (2.98)$$

Again we see our old friend f_T and γ playing a central role.

Phase noise

Owing to random noise from devices, an oscillator does not have a delta-function power spectrum, but rather has a very sharp peak at the oscillation frequency (Figure 2.35). The amplitude of the noise power spectrum drops very quickly, though, as one moves away from the center frequency. For example, a cell phone oscillator has a phase noise that is 100 dB down at an offset of only 0.01% from the carrier! Note that the noise shaping is not due to the LC tank or crystal resonator used to build an oscillator, which has a much wider bandwidth.

The importance of phase noise is that it places limits on RF communication. For example, phase noise in a transmit chain will "leak" power into adjacent channels as shown in Figure 2.36. Since the power transmitted is large—say, about

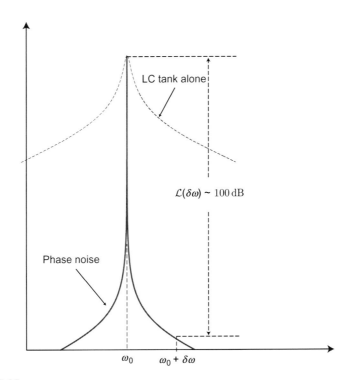

FIGURE 2.35

The power spectrum of an oscillator has a finite width in the frequency domain owing to the noise in electronic devices which alters the period of oscillation (jitter).

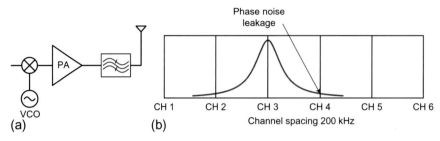

(a)   (b)

FIGURE 2.36

(a) In a typical up-conversion transmitter, the "local oscillator" has phase noise and this is amplified by the power amplifier driving the antenna. (b) This phase noise leaks into adjacent channels, making it impossible or difficult for receivers to distinguish the information from the noise.

30 dBm—an adjacent channel in a narrowband system may reside only about 200 kHz away,[2] placing a stringent specification on the transmitter spectrum.

In a receive chain, the fact that the local oscillator is not a perfect delta function means that there is a continuum of local oscillators that can mix with interfering signals and produce energy at the same intermediate frequency. Here, we observe an adjacent channel signal mixing with the "skirt" of the local oscillator and falling on top of the a weak intermediate-frequency signal from the desired channel.

In a digital communication system, phase noise can lead to a lower noise margin. In Figure 2.37, we see that the phase noise causes the constellation of a quadrature phase-shift keying modulation to spread out. In orthogonal frequency-division multiplexing systems, a wide bandwidth is split into subchannels. The phase

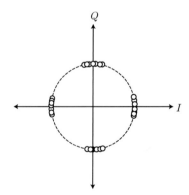

FIGURE 2.37

The ideal constellation of a transmitter and receiver is degraded by the "jitter" or phase noise in the RF carriers and clocks used to modulate and sample the data.

---

[2]In fact, these are the specifications for the popular 2G digital mobile standard known as GSM.

noise leads to intercarrier interference and a degradation in the digital communication bit error rate.

Phase noise derivation: Lorentzian spectrum

While the full derivation of phase noise requires the solving of stochastic differential equations, we can gain insight with a very simple linear time-invariant model. It can be readily shown that the power spectrum of v_o is given by [16, 17]

$$v_o^2 = \overline{i_n^2} \frac{R_1^2}{(1 - (g_m R_1/n))^2 + 4Q^2(\delta\omega^2/\omega_0^2)}. \tag{2.99}$$

Thus, the noise has a Lorentzian shape for white noise. For offset frequencies of interest,

$$4Q^2 \frac{\delta\omega^2}{\omega_0^2} \gg \left(1 - \frac{g_m R_1}{n}\right)^2. \tag{2.100}$$

The spectrum normalized to the peak is given by

$$\left(\frac{v_o}{V_o}\right)^2 \approx \frac{\overline{i_n^2} R_1^2}{V_o^2} \left(\frac{\omega_0}{\delta\omega}\right)^2 \frac{1}{4Q^2}. \tag{2.101}$$

The above equation is in the form of Leeson's equation. It compactly expresses that the oscillator noise is expressed as noise power over signal power (N/S), divided by Q^2 and drops like $1/\delta\omega^2$.

Phase noise and flicker noise

As we have seen, phase noise depends on the device noise, which is similar to our analysis of linear gain. Namely, phase noise at some frequency offset is generated by device noise at the same frequency. But in our simple linear time-invariant model, we disregarded the important quasi-periodically time-varying nature of the transistor operating point. As shown in Figure 2.38, this time-varying behavior, similar to that

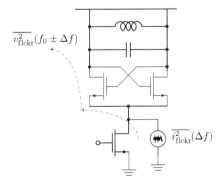

FIGURE 2.38

The noise from low frequencies, such as flicker noise in the current source, is up-converted to RF through the time-varying "mixer" action of the oscillator.

of a mixer, allows noise from low frequency to mix up and appear around the RF carrier. Thus, in RF circuits, noise at low frequency (due to flicker) is also important and cannot be disregarded.

2.4.5 LINEARITY

At RF frequencies, in addition to HD metrics, there is also considerable interest in intermodulation distortion, especially odd-order intermodulation. The reason for this is that many RF systems are narrowband and thus a bandpass network can attenuate distortion products that fall out of the desired band, such as harmonics, but odd-order intermodulation products fall into the band and cannot be filtered. Moreover, even-order intermodulation products have DC components that are important in direct-conversion applications, since most of the signal processing is at the baseband (DC).

To derive the intermodulation of a general nonlinearity, we can follow the same approach as before. Consider applying a two-tone signal to a memoryless nonlinearity described by a power series:

$$i_{ds} = g_{m1}(V_{gs1}\cos(\omega_1 t) + V_{gs2}\cos(\omega_2 t)) + g_{m2}(V_{gs1}\cos(\omega_1 t) + V_{gs2}\cos(\omega_2 t))^2 + \cdots.$$
$$(2.102)$$

If we focus only on the second power term, we see that in addition to harmonics at $2\omega_{1,2}$ we also generate a new second-order intermodulation (IM$_2$) term,

$$g_{m2}(V_{gs1}\cos(\omega_1 t) + V_{gs2}\cos(\omega_2 t))^2 = g_{m2}V_{gs,1,2}^2 \frac{1 + \cos(2\omega_{1,2})}{2}$$
$$+ 2V_{gs1}\cos(\omega_1 t)V_{gs2}\cos(\omega_2 t), \qquad (2.103)$$

where the latter term can be simplified to

$$= \cdots + g_{m2}V_{gs1}V_{gs2}(\cos(\omega_1 + \omega_2)t + \cos(\omega_1 - \omega_2)t). \qquad (2.104)$$

These two "sum and difference" frequency terms are the second-order intermodulation terms. The ratio of their power relative to the fundamental is used to define the IM$_2$ metric (by applying equal voltages for both tones):

$$\mathrm{IM}_2 = \frac{g_{m2}V_{gs,1}^2}{g_{m1}V_{gs}} = \frac{g_{m2}}{g_{m1}}V_{gs}. \qquad (2.105)$$

We see that for a memoryless system, IM$_2$ is in fact related to HD$_2$ (see Equation 2.26) as follows:

$$\mathrm{IM}_2 = 2\mathrm{HD}_2. \qquad (2.106)$$

This relationship clearly breaks down if there is memory in the system. A simple low-pass filter will attenuate HD$_2$ but leave the low-frequency difference tone at $\omega_1 - \omega_2$ intact.

In a similar way, we can proceed to analyze the third-order two-tone response, building from here

$$= \cdots + g_{m3}(V_{gs1}\cos(\omega_1 t) + V_{gs2}\cos(\omega_2 t))^3 \tag{2.107}$$

$$= \cdots + g_{m3}(V_{gs1}\cos(\omega_1 t) + V_{gs2}\cos(\omega_2 t))^2(V_{gs1}\cos(\omega_1 t) + V_{gs2}\cos(\omega_2 t)) \tag{2.108}$$

$$= \cdots + g_{m3}V_{gs1}V_{gs2}\left[\frac{1}{2}(\cos(2\omega_{1,2}) + 1)\right.$$

$$\left. + \cos(\omega_1 + \omega_2)t + \cos(\omega_1 - \omega_2)t\right](V_{gs1}\cos(\omega_1 t) + V_{gs2}\cos(\omega_2 t)). \tag{2.109}$$

Expanding the products, we see terms that now have terms at $2\omega_2 + \omega_1$ and $2\omega_1 + \omega_2$, generated by

$$\left(\frac{1}{2} + \frac{1}{4}\right)\cos(\omega_1 + \omega_2)t \times \cos(\omega_{1,2}t), \tag{2.110}$$

and also terms at $2\omega_2 - \omega_1$ and $2\omega_1 - \omega_2$, generated by

$$\left(\frac{1}{2} + \frac{1}{4}\right)\cos(\omega_1 - \omega_2)t \times \cos(\omega_{1,2}t). \tag{2.111}$$

These are called the third-order intermodulation terms, or IM_3. Computing the ratio of the third-order intermodulation to the fundamental (with equal powers in each tone), we have

$$IM_3 = \frac{g_{m3}V_{gs}^3(3/4)}{g_{m1}V_{gs}} = \frac{3}{4}\frac{g_{m3}}{g_{m1}}V_{gs}^2. \tag{2.112}$$

We see that IM_3 also has many of the same features as HD_3; in fact, for a memoryless system they are related by a factor of 3. What is notably different about IM_3 (and in fact all odd-order intermodulation products) is that there are spectral tones created in band, or near the original two-tone frequencies. In particular, these tones are at $2\omega_2 - \omega_1$ and $2\omega_1 - \omega_2$, and as $\omega_1 \approx \omega_2$ in a narrowband system, these distortion products fall in band and can do the most harm. It is for this reason that IM_3 plays such an important role in RF systems.

Figure 2.39 displays the output signal at the fundamental versus the distortion products at frequencies $\omega_2 - \omega_1$ (second order) and $2\omega_2 - \omega_1$ (third order). The distance between the fundamental curve and the second- and third-order curves is the intermodulation on a log scale. The intercept (extrapolated) is another metric that is often used to describe the third- and second-order intermodulation with a single number. By definition, at the intercept, V_{IIP} is the input signal level such that the distortion product and the fundamental output have equal magnitude.

IM_3 and IM_2 are plotted versus the signal power in Figure 2.40. These plots are very similar to the HD plots and have a slope of 2 and 1, respectively, on a decibel scale. Compared with the previous plots, the slopes are 1 minus the slope of the power that generates the nonlinearity, since IM_2 and IM_3 are ratios normalized by the fundamental.

For the second-order distortion products,

$$IM_2 = 1 = \frac{g_{m2}}{g_{m1}}V_{IIP,2}, \tag{2.113}$$

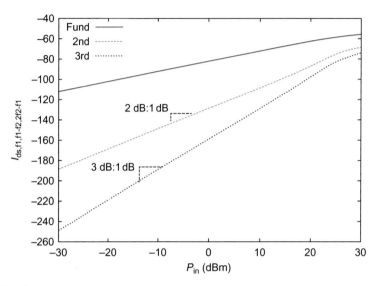

FIGURE 2.39

The power of fundamental and the second and third intermodulation products as a function of signal power. The intermodulation products have a slope of 2 and 3 dB per decibel increase in signal power and the distance to the fundamental is the intermodulation ratio on a log scale.

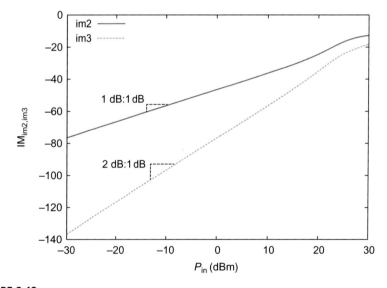

FIGURE 2.40

IM_2 and IM_3 as a function of signal power have a slope of 1 and 2 dB per decibel increase in signal power, respectively.

$$V_{\text{IIP},2} = \frac{g_{m1}}{g_{m3}}, \tag{2.114}$$

For the third-order distortion products,

$$\text{IM}_3 = 1 = \frac{3}{4}\frac{g_{m3}}{g_{m1}}V_{\text{IIP},3}^2, \tag{2.115}$$

$$V_{\text{IIP},3} = \sqrt{\frac{4}{3}\frac{g_{m1}}{g_{m3}}}. \tag{2.116}$$

The advantage of the intercept is that we can calculate (or measure) it once, and then for any given input signal level power, we can simply apply the rule that the IM_3 relation grows by 2 dB for every 1 dB of back-off from the intercept, whereas the IM_2 relation grows decibel for decibel.

Memory effects

As previously mentioned, for RF systems we expect that any reactance in the circuit, such as device capacitors, will interact with the distortion products generated by the transistor nonlinearity and attenuate (and perhaps magnify) the distortion products. This extra complication is handled by using a Volterra series [1] in place of a power series. Moreover, any nonlinearity in the capacitors (such as the variation of C_{gs} with signal power) will also create distortion products. This means that the *I-V* and *C-V* relations both need to be extracted accurately for RF applications. From a compact modeling perspective, a simple solution is to focus on the *C-V* curve of the device and ensure a good fit. Similarly to the case of the *I-V* curve, the fit needs to be very good in the sense that not only must the absolute values fit the measurements, but the curves must be smooth and symmetric since distortion products are created by the derivatives of the *C-V* curve. Owing to space limitations, the details of this are omitted here.

Other distortion metrics

While IM_2 and IM_3 are excellent vehicles for understanding the distortion generation in narrowband systems, when an actual modulation is applied to a transistor, all intermodulation products produce distortion, causing the spectrum of the signal to expand. This is called spectral regrowth, and in addition to corruption of the modulation pattern, it also introduces unwanted out-of-band interference, in particular to neighboring channels (in a frequency-division multiplexed system). The distortion in the modulation pattern is measured by the error vector magnitude, which measures the deviation from the ideal constellation points due to the nonlinearity (and noise). As these metrics are more applicable at the system level, rather than at the device level, they are hard to use directly in compact modeling. In practice, if intermodulation products match up to several orders, this is a good sign that the compact model is representing the distortion products well. In particular, the intermodulation products must also match for different tone spacings (even though simple theory predicts the intermodulation products do not depend on tone spacing, the more detailed Volterra series analysis will show otherwise). Furthermore, three or more tones should be used to emulate the effect of a broadband modulation, such as an orthogonal frequency-division multiplexing or multitone signal.

2.5 CONCLUSION

Today, compact models play a key binding role between technology and circuits, for diverse applications including precision analog, digital, mixed signal, and RF and millimeter-wave applications. These vast applications place stringent demands on the accuracy of the compact model, requiring extensive testing and validation. In this chapter, we have examined important properties of a compact model for both analog and RF applications. We have seen that common metrics such as threshold voltage variation, intrinsic gain, device unity gain f_T, maximum oscillation frequency f_{max}, minimum achievable noise figure F_{min}, and Mason's unilateral gain U all play an important role in evaluating the accuracy of a compact model. We have also seen that other metrics and properties that capture linearity, device variability, symmetry, and other subtler factors are also important. Many subsequent chapters will focus on compact model modules that directly impact these metrics.

REFERENCES

[1] W.J. Rugh, Nonlinear System Theory: The Volterra-Wiener Approach, Johns Hopkins University Press, Baltimore, 1981.

[2] A.M. Niknejad, Electromagnetics for High-Speed Analog and Digital Communication Circuits, Cambridge University Press, Cambridge, 2007.

[3] C. Doan, S. Emami, A. Niknejad, R. Brodersen, Millimeter-wave CMOS design, IEEE J. Solid State Circuits 40 (1) (2005) 144–155.

[4] International Technology Roadmap for Semiconductor, RF and Analog/Mixed-Signal Technologies (RFAMS). Available from: http://www.itrs.net/.

[5] Z. Deng, A. Niknejad, A layout-based optimal neutralization technique for mm-wave differential amplifiers, in: IEEE Radio Frequency Integrated Circuits Symposium (RFIC), May 23–25, 2010, pp. 355–358.

[6] S.J. Mason, Power gain in feedback amplifiers, Trans. IRE Professional Group Circuit Theory CT-1 (2) (1954) 20–25.

[7] M.S. Gupta, Power gain in feedback amplifiers, a classic revisited, IEEE Trans. Microw. Theory Tech. 40 (5) (1992) 864–879.

[8] S.V. Thyagarajan, A.M. Niknejad, Manuscript in preparation.

[9] S. Venugopalan, From Poisson to silicon—advancing compact SPICE models for IC design (Ph.D. dissertation), Electrical Engineering and Computer Sciences, University of California at Berkeley, Technical Report No. UCB/EECS-2013-166, http://www.eecs.berkeley.edu/Pubs/TechRpts/2013/EECS-2013-166.html.

[10] M. Chan, K. Hui, R. Neff, C. Hu, P.K. Ko, A relaxation time approach to model the non-quasi-static transient effects in MOSFETs, in: International Electron Devices Meeting, Technical Digest, December 1994, pp. 169–172.

[11] X. Jin, J.-J. Ou, C.-H. Chen, W. Liu, M.J. Deen, P.R. Gray, C. Hu, An effective gate resistance model for CMOS RF and noise modeling, in: International Electron Devices Meeting (IEDM), Technical Digest, December 1998, pp. 961, 964.

[12] A.J. Scholten, R. van Langevelde, L.F. Tiemeijer, R.J. Havens, D.B.M. Klaassen, Compact MOS modelling for RF CMOS circuit simulation, Simulation of Semiconductor Processes and Devices, Springer, Vienna, 2001, pp. 194–201.

[13] H. Wang, T.-L. Chen, G. Gildenblat, Quasi-static and nonquasi-static compact MOSFET models based on symmetric linearization of the bulk and inversion charges, IEEE Trans. Electron Devices 50 (11) (2003) 2262–2272.

[14] D.M. Pozar, Microwave Engineering, second ed., Wiley, New York, NY, 1997.

[15] M.W. Pospieszalski, On the measurement of noise parameters of microwave two-ports, IEEE MTT-S Int. Microw. Symp. Digest 34 (4) (1986) 456–458.

[16] D. Leeson, A simple model of feedback oscillator noise spectrum, Proc. IEEE 54 (2) (1966) 329–330.

[17] A.M. Niknejad, R.G. Meyer, Design, Simulation and Applications of Inductors and Transformers for Si RF ICs, Kluwer Academic Publishers, Boston, 2000.

Core model for FinFETs

3

CHAPTER OUTLINE

3.1 Core model for double-gate FinFETs 72
3.2 Unified FinFET compact model 80
Chapter 3 Appendix: Explicit surface potential model 87
3A.1 Continuous starting function 88
3A.2 Quartic modified iteration: Implementation and evaluation 91
References 96

In the implementation of circuit simulators, compact models are preferred over other numerical approaches because the former can offer, in addition to good computational efficiency, good accuracy [1]. The good accuracy of a compact model mainly lies in the amount of physics and the assumptions behind its mathematical derivation. Indeed, FinFETs are constructed in the nanoscale regime; therefore, accurate compact models must include several physical effects: charge quantization, gate oxide tunneling, gate capacitance degradation, short-channel effects, etc.

Most of the compact models are based on a "core model," which is a model obtained using a long-channel assumption, the so-called gradual-channel approximation (GCA) [2], and simplifying other physical effects such as charge quantization or gate oxide tunneling. A core model is crucial for the completed compact model because it gives the basis of a mathematical framework which is continuous and smooth, a key requirement for a robust model. In this context, core models are further improved by the inclusion of correction terms that represent advanced physical effects [1, 3]. Regularly, core models are obtained by solving Poisson's equation under the GCA condition and assuming Boltzmann statistics for the carriers. Even though the use of the GCA condition and Boltzmann statistics alleviates the difficulty in obtaining a solution from Poisson's equation, a direct analytical solution is available only for the cases of undoped double-gate (DG) [4] and cylindrical gate-all-around (Cy-GAA) FETs [5], where the three-dimensional Poisson's equation can be reduced to a one-dimensional form. If depletion charges arising from dopants are included, Poisson's equation becomes highly nonlinear. It is then more challenging to obtain a direct analytical solution [3, 6–9]. However, in realistic FinFETs, doping is needed to be used for multiple-threshold-voltage devices that are required in contemporary system-on-chip technologies for better power performance-area trade-off [10]; thus, core models including doping effects must be developed. Finding a direct analytical solution becomes even more difficult for complex FinFET geometries because they lack structural symmetry [11]. Indeed, compact models for asymmetric geometries, such as triple-gate (TG), rectangular GAA (Re-GAA), or pi-gate FETs, are rarely found in the literature and are accomplished only by the extensive use of fitting parameters or numerical techniques [11–14]. However, some of these asymmetric geometries offer simpler fabrication processes than other symmetric geometries [15]. Therefore, it is important to develop a physical-based core model for FinFETs with complex geometry, for comprehensive understanding and circuit design.

Various compact models have been proposed for several FinFET geometries, such as DG [3, 4, 6–8, 16–19], TG [11, 12, 20], Re-GAA [11, 13, 14], and Cy-GAA [5, 9, 21, 22]. The core model of BSIM-CMG is a surface potential model which is based on the solution of Poisson's equation for a doped DG FinFET. The surface potential obtained at the channel is used to obtain the mobile charge in the channel, which eventually leads to the computation of the drain current of the device. A description of this model is given in Section 3.1. Recently, a new core model has been added to BSIM-CMG, and it is entitled unified FinFET compact model owing its capability to describe the electrical behavior of several FinFETs with complex cross-sections [23]. This new feature of BSIM-CMG is presented in Section 3.2.

3.1 CORE MODEL FOR DOUBLE-GATE FinFETs

The core model used in BSIM-CMG is based on a solution of Poisson's equation for a long-channel DG FinFET, assuming a finite doping in the channel to mimic the doped channels currently used in FinFET fabrication [10]. It is challenging to obtain a direct analytical solution of Poisson's equation for doped FinFETs owing to the high nonlinearity of the equation; therefore, to overcome this limitation, a perturbation approach is used to approximately solve Poisson's equation in the presence of body doping [3, 6].

Figure 3.1 shows a two-dimensional cross-section of a DG FinFET which is being used as a reference for the model derivation. Poisson's equation, assuming the GCA and the Boltzmann distribution for the inversion carriers, and considering only mobile carriers (e.g., electrons in an n-type MOS FinFET), can be expressed as

$$\frac{\partial^2 \psi(x, y)}{\partial x^2} = \frac{q}{\varepsilon_{ch}} \left(n_i e^{\frac{\psi(x,y) - \psi_B - V_{ch}(y)}{V_{tm}}} + N_{ch} \right), \tag{3.1}$$

where $\psi(x, y)$ is the electrostatic potential in the channel, q is the magnitude of the electronic charge, n_i is the intrinsic carrier concentration, ε_{ch} is the dielectric

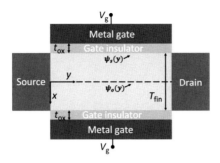

FIGURE 3.1

A symmetric DG FinFET.

constant of the channel (fin), V_{tm} is the thermal voltage given by $k_B T/q$, where k_B and T are the Boltzmann constant and the temperature, respectively, V_{ch} is the quasi-Fermi potential of the channel ($V_{ch}(0) = V_s$ and $V_{ch}(L) = V_d$), which only has a y spatial dependence, N_{ch} is the channel doping, and $\psi_B = V_{tm} \ln(N_{ch}/n_i)$. Note that (Equation 3.1) is a one-dimensional Poisson equation; the other two dimensions have been disregarded owing to the long length of the channel (GCA condition) and the symmetry of the DG structure in the z-direction. In other FinFET geometries, where the channel length is short and the fin cross-sections are complex, the full three-dimensional Poisson equation must be solved.

Figures 3.2 and 3.4 show the fin potential versus fin position obtained from Equation (3.1) for different conditions. Equation (3.1) has been solved numerically using the finite element method to generate the fin potential data. From these results, there are several points that must be contrasted with conventional planar MOSFETs. For example, as shown in Figures 3.2 and 3.3, the center potential is not fixed as in the case of the bulk potential in planar MOSFETs. Indeed, the center potential has a different value depending on the bias condition, and it tends to a fixed value in the strong operation regime. In addition, FinFETs may not require the use of a high doping concentration for the channel to counter short-channel effects. In this context, lightly doped fins consequently increase carrier mobility and reduce device variability coming from random dopant fluctuations. The use of lightly doped channels implies that the potential in the subthreshold region is mostly flat, as shown in Figure 3.2, which makes the mobile carries and subthreshold current be proportional to the fin thickness. Thus, in order to decrease leakage, the fin thickness must be scaled down. Including dopants in the channel can also be a good option for multiple-threshold FinFETs [10]. Figure 3.3 shows how the potential changes when a heavily doped fin is used. The potential in the subthreshold region is bent owing the ionized dopants, which change the threshold of the device depending on the amount of doping and the thickness of the fin. Finally, another important point to notice is that as thickness of the fin is decreased, the potential in the center of the channel

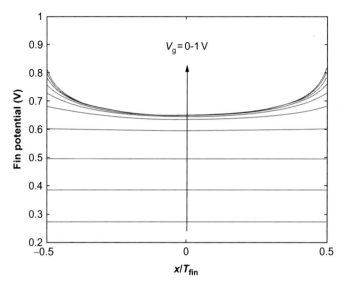

FIGURE 3.2

Fin potential versus position obtained from the numerical solution of Equation (3.1) for a lightly doped fin. $N_{ch} = 1 \times 10^{15}\,cm^{-3}$, $T_{fin} = 20\,nm$, $t_{ox} = 1\,nm$, and $V_{ch} = 0\,V$ were used for the simulation.

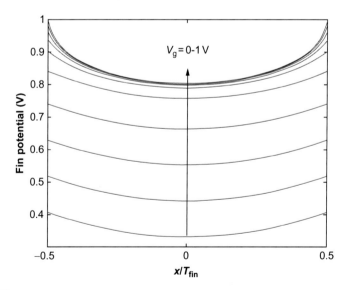

FIGURE 3.3

Fin potential versus position obtained from the numerical solution of Equation (3.1) for a doped fin. $N_{ch} = 5 \times 10^{18}\,cm^{-3}$, $T_{fin} = 20\,nm$, $t_{ox} = 1\,nm$, and $V_{ch} = 0\,V$ were used for the simulation.

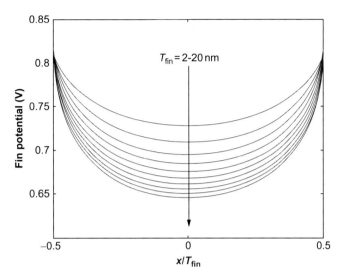

FIGURE 3.4

Fin potential versus position obtained from the numerical solution of Equation (3.1) for different fin thicknesses at strong-inversion bias. $N_{ch} = 1 \times 10^{15}\,cm^{-3}$, $V_{ch} = 0\,V$, $T_{fin} = 20\,nm$, $t_{ox} = 1\,nm$, and $V_g = 1\,V$ were used for the simulation. As T_{fin} decreases, the center potential increases, which increases the number of mobile carriers in the center of the fin.

increases as shown in Figure 3.3. An increase of the center potential increases the number of mobile carriers in the center of the fin, where mobility is greater than at the surface, and thus higher mobility can be obtained for those carriers. Figures 3.2 and 3.3 are obtained by a numerical solution which is not suitable for a compact modeling perspective; therefore, a compact model will be presented starting from Equation (3.1).

In order to obtain the potential in the channel for Equation (3.1), ψ is written following the perturbation approach [3]:

$$\psi\,(x, y) \cong \psi_1\,(x, y) + \psi_2\,(x, y). \tag{3.2}$$

Here, ψ_1 is the potential contribution due to the inversion carriers and without the effect of the ionized dopants, N_{ch}, and is given by

$$\frac{\partial^2 \psi_1(x, y)}{\partial x^2} = \frac{q n_i}{\varepsilon_{ch}} e^{\frac{\psi_1(x,y) - \psi_B - V_{ch}(y)}{V_{tm}}}, \tag{3.3}$$

and ψ_2 is the potential contribution due to the presence of the ionized dopants N_{ch}, and without the effect of the inversion carriers; it is the perturbation potential and is given by

$$\frac{\partial^2 \psi_2(x, y)}{\partial x^2} = \frac{qN_{ch}}{\varepsilon_{ch}}.$$ (3.4)

The geometrical symmetry of a DG FinFET leads to the fact that the vertical component of the electric field $\mathcal{E}_x$ at the center of the channel is zero, and then it is possible to integrate Equation (3.3) twice to obtain $\psi_1(x, y)$ as a function of the potential in the center of the body $\psi_0(y)$:

$$\psi_1(x, y) = \psi_0(y) - 2V_{tm}\ln\left[\cos\left(\sqrt{\frac{q}{2\varepsilon_{ch}V_{tm}}\frac{n_i^2}{N_{ch}}e^{\frac{\psi_0(y)-V_{ch}(y)}{V_{tm}}}} \times \frac{x}{2}\right)\right].$$ (3.5)

In order to find ψ_2, it is also possible to apply $\mathcal{E}_x = 0$ at the center of the channel and set $\psi_2(x = 0, y) = 0$. Then, integrating Equation (3.4) twice, one can obtain

$$\psi_2(x, y) = \frac{qN_{ch}x^2}{2\varepsilon_{ch}}.$$ (3.6)

The surface potential ψ_s at any point y along the surface is obtained by evaluating the sum of ψ_1 and ψ_2 at the surface of the fin:

$$\psi_s(y) \cong \psi_1(-T_{fin}/2, y) + \psi_2(-T_{fin}/2, y).$$ (3.7)

Gauss's law and the boundary conditions at the channel-insulator interface lead to a second important equation:

$$V_{gs} = V_{fb} + \psi_s(y) + \varepsilon_{ch}\mathcal{E}_{xs}/C_{ox},$$ (3.8)

where V_{gs} is the gate voltage, V_{fb} is the flat-band voltage, C_{ox} is the gate oxide capacitance per unit area, given by ε_{ox}/T_{ox}, where ε_{ox} and T_{ox} are the oxide dielectric constant and oxide thickness, respectively, and $\mathcal{E}_{xs}$ is the vertical component of the electric field at the surface, which can be obtained by integrating Equation (3.1),

$$\mathcal{E}_{xs} = \sqrt{\frac{2qn_i}{\varepsilon_{ch}}\left[V_{tm}\left(e^{\frac{\psi_s(y)}{V_{tm}}} - e^{\frac{\psi_0(y)}{V_{tm}}}\right)e^{\frac{-\psi_B - V_{ch}(y)}{V_{tm}}} + e^{\frac{\psi_B}{V_{tm}}}(\psi_s(y) - \psi_0(y))\right]}.$$ (3.9)

Equations (3.7) and (3.8) represent a self-consistent system of equations that can be used to obtain ψ_0 and ψ_s. However, through a change of variable given by

$$\beta = \sqrt{\frac{q}{2\epsilon_{ch}V_{tm}}\frac{n_i^2}{N_{ch}}e^{\frac{\psi_0-V_{ch}}{V_{tm}}}}\frac{T_{fin}}{2},$$ (3.10)

Equations (3.7) and (3.8) can be written as a single equation:

$$f(\beta) \equiv \ln\beta - \ln(\cos\beta) - \frac{V_{gs} - V_{fb} - V_{ch}}{2V_{tm}} + \ln\left(\frac{2}{T_{fin}}\sqrt{\frac{2\varepsilon_{ch}V_{tm}N_{ch}}{qn_i^2}}\right)$$

$$+ \frac{2\varepsilon_{ch}}{T_{fin}C_{ox}}\sqrt{\beta^2\left(\frac{e^{\frac{\psi_{pert}}{V_{tm}}}}{\cos^2\beta} - 1\right)} + \frac{\psi_{pert}}{V_{tm}^2}\left[\psi_{pert} - 2V_{tm}\ln(\cos\beta)\right] = 0, \quad (3.11)$$

where ψ_{pert} is given by ψ_2 evaluated at $x = T_{fin}/2$. Equation (3.11) is an implicit equation in β which must be solved using numerical methods. Then, once β has been calculated, the surface potential and the charge in the channel can be obtained. Figure 3.5 shows the surface potential obtained from Equation (3.11) and the numerical solution of Equation (3.1) for different doping concentrations. The amount of doping in the channel determines the threshold voltage of the device as shown in Figure 3.6, which represents the mobile charge density obtained from the proposed compact model and the numerical solution of Figure 3.6 for different doping concentrations. In the case of lightly doped DG FinFETs, the thickness of the channel determines the mobile carrier charge density in the channel in a linear manner as shown in Figure 3.7.

Solving Equation (3.11) using numerical methods is not practical for compact modeling applications because the use of them increases the computation time and may cause divergence problems [24]. Therefore, Equation (3.11) is solved by first using an analytical approximation for the initial guess, followed by two quartic

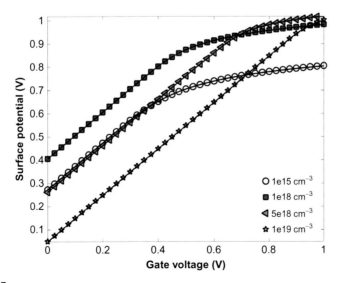

FIGURE 3.5

Surface potential versus V_g of DG FinFETs ($T_{fin} = 20\,\text{nm}$) at $V_{ds} = 0.0\,\text{V}$ obtained from the proposed model (lines) and numerical simulations (symbols) for different channel doping.

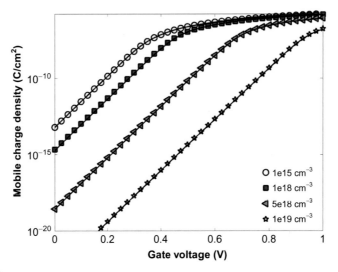

FIGURE 3.6

Mobile electron charge density versus V_g of DG FinFETs ($T_{fin} = 20\,nm$) at $V_{ds} = 0.0\,V$ obtained from the proposed model (lines) and numerical simulations (symbols) for different channel doping.

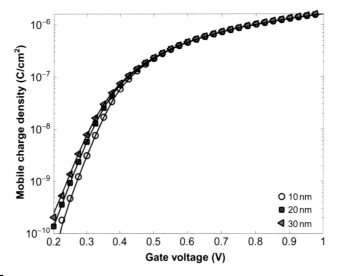

FIGURE 3.7

Mobile electron charge density versus V_g of DG FinFETs (using an intrinsic fin channel) at $V_{ds} = 0.0\,V$ obtained from the proposed model (lines) and numerical simulations (symbols) for different fin thicknesses.

modified iterations [32]. This approach makes the model numerically robust and accurate (see Section 3.2). The surface potentials at the source end ψ_s and drain end ψ_d are calculated by setting $V_{ch} = V_s$ and $V_{ch} = V_d$, respectively. For a lightly doped body, Equation (3.11) can be simplified further [3] to speed up the simulation. This option can be selected in BSIM-CMG by setting the parameter COREMOD. A separate model has been derived for the cylindrical gate geometry, which has been discussed in detail in [22].

In order to complete the core model, the drain-to-source current I_{ds} model in BSIM-CMG is obtained from a solution of the drift-diffusion equation, assuming a long-channel DG FinFET:

$$I_{ds}(y) = \mu(T)WQ_{inv}(y)\frac{dV_{ch}}{dy}, \tag{3.12}$$

where $\mu(T)$ is the low-field and temperature-dependent mobility, W is the total effective width, and Q_{inv} is the inversion charge per unit area in the body. Equation (3.12) includes drift and diffusion transport mechanisms through the use of the quasi-Fermi potential.

Integrating both sides of Equation (3.12), and considering the fact that under quasi-static operation I_{ds} is constant along the channel, one can express Equation (3.12) in its integral form:

$$I_{ds} = \frac{W}{L}\mu(T)\int_{Q_{inv,s}}^{Q_{inv,d}} Q_{inv}\left(\frac{dV_{ch}}{dQ_{inv}}\right)dQ_{inv}, \tag{3.13}$$

where L is the effective channel length, and $Q_{inv,s}$ and $Q_{inv,d}$ are the inversion charge densities at the source and drain ends, respectively, given by

$$Q_{inv,d/s} = C_{ox}(V_{gs} - V_{fb} - \psi_{d/s}) - Q_{bulk}. \tag{3.14}$$

Here, Q_{bulk} is the fixed depletion charge and is given by $qN_{ch}T_{fin}$ and $\psi_{d/s}$ are obtained by solving Equation (3.11). The term dV_{ch}/dQ_{inv} in Equation (3.13) can be calculated as a function of Q_{inv} using a simple, but accurate, implicit equation for Q_{inv} [3]:

$$Q_{inv}(y) \approx \sqrt{2qn_i\varepsilon_{ch}V_{tm}}\, e^{\frac{\psi_s(y)-\psi_B-V_{ch}(y)}{2V_{tm}}}\sqrt{\frac{Q_{inv}(y)}{Q_{inv}(y)+Q_0}}, \tag{3.15}$$

where $Q_0 = Q_{bulk} + 5C_{fin}V_{tm}$, with $C_{fin} = \varepsilon_{ch}/T_{fin}$. With use of this approximation, Equation (3.13) can be integrated analytically, leading to the following basic equation for I_{ds}:

$$I_{ds} = \mu(T)\cdot\frac{W}{L}\cdot\left[\frac{Q_{inv,s}^2 - Q_{inv,d}^2}{2C_{ox}} + 2V_{tm}(Q_{inv,s} - Q_{inv,d}) - V_{tm}Q_0\ln\left(\frac{Q_0 + Q_{inv,s}}{Q_0 + Q_{inv,d}}\right)\right]. \tag{3.16}$$

Note that Q_{inv} charges are calculated using Equations (3.11) and (3.13). Figure 3.8 shows an example drain current obtained from the proposed model and numerical

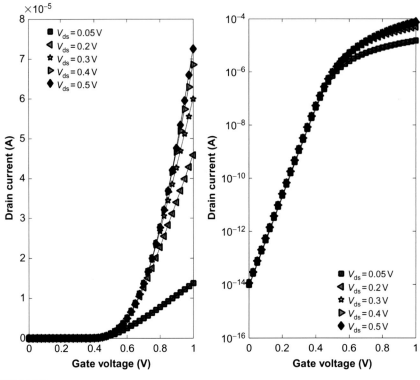

FIGURE 3.8

Drain current versus V_g of DG FinFETs (using an intrinsic fin channel) at different V_{ds} values obtained from the proposed model (lines) and numerical simulations (symbols). Linear (left) and logarithmic (right) scales are shown. $H_{fin} = L = 1\,\mu m$, $t_{ox} = 1\,nm$, $N_{ch} = 1 \times 10^{18}\,cm^{-3}$, $\mu_e = 100\,cm^2\,V^{-1}\,s^{-1}$, and a gate work function equal to 4.6 eV were used for the simulations.

simulations. It shows that BSIM-CMG captures the behavior of a FinFET under all bias conditions: subthreshold, triode, and saturation conditions.

3.2 UNIFIED FINFET COMPACT MODEL

The typical rectangular cross-section of FinFETs is hardly found in industry FinFETs. Indeed, whether intentional or due to manufacturing variation, industry FinFET cross-sections are nonuniform and similar to rounded trapezoidal shapes [25, 26], as shown in Figure 3.9. In order to capture fin shape effects on device performance, a compact model for FinFETs with complex cross-sections is important. In this section, the unified FinFET compact model is presented for devices with complex fin

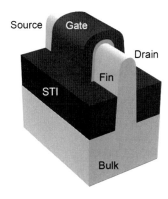

FIGURE 3.9

A three-dimensional schematic of a FinFET with a complex fin cross-section. The fin shape is similar to that of industry FinFETs reported in [25, 26].

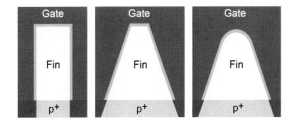

FIGURE 3.10

Cross-section schematics of FinFETs with complex fin shapes. The left structure represents a perfectly rectangular fin shape, and the two subsequent structures represent a trapezoidal fin shape with and without rounded corners. These structures may differ from the typical rectangular FinFET shape because of an intentional manufacturing process or random structure variation. The electrical behavior of each of these devices can be obtained with the proposed unified compact model.

cross-sections. FinFET structures such as DG, Cy-GAA, Re-GAA, and rounded trapezoidal TG FinFETs are all modeled under the same framework. A single unified core model is used for different FinFET structures as shown in Figure 3.10, and only model parameters are different for each FinFET structure, which are precalculated for each device type and dimension. The proposed core model can be used with short-channel-effect submodels in a similar manner as was done in previous versions of BSIM-CMG models [27]. The model presented in this section has been recently being incorporated to BSIM-CMG [23].

Several compact models have been proposed for FinFETs with complex cross-sectional shapes. The work presented in [11] developed compact models for different shapes of undoped or lightly doped FinFETs using a combination of the compact models for DG [4] and Cy-GAA [5] FinFETs. In [28], a compact model for undoped or lightly doped FinFETs was extended to model FinFETs with different cross-sectional shapes by obtaining an equivalent channel thickness for each structure. Another compact model has recently been proposed for FinFET devices with different cross-sectional shapes [29, 30], where new models for doped FinFETs were developed in a universal model framework. In this section, on the basis of the approach presented in [23], a new normalized unified FinFET core model is presented [23]. The new normalized charge model is obtained from the solutions of Poisson's equation for DG and Cy-GAA FinFETs, which leads to a single closed-form relationship between the mobile charge and the applied terminal voltages given as follows [23]:

$$v_g - v_o - v_{ch} = -q_m + \ln(-q_m) + \ln\left(\frac{q_t^2}{e^{q_t} - q_t - 1}\right),$$ (3.17)

where

$$v_o = v_{fb} - q_{dep} - \ln\left(\frac{2qn_i^2 A_{ch}}{v_T C_{ins} N_{ch}}\right)$$ (3.18)

and

$$q_t = (q_m + q_{dep})r_N.$$ (3.19)

In the previous equations, v_g and v_{ch} are the normalized gate and channel potentials expressed by

$$v_g = \frac{V_g}{v_T}$$ (3.20)

and

$$v_{ch} = \frac{V_{ch}}{v_T}.$$ (3.21)

q_m and q_{dep} are the normalized mobile and depletion charges:

$$q_m = \frac{Q_m}{v_T C_{ins}},$$ (3.22)

$$q_{dep} = \frac{-qN_{ch}A_{ch}}{v_T C_{ins}},$$ (3.23)

r_N is given by

$$r_N = \frac{A_{fin}C_{ins}}{\varepsilon_{ch}W^2},$$ (3.24)

where A_{ch} is the area of the channel, N_{ch} is the doping in the channel, W is the channel width, and C_{ins} is the insulator capacitance per unit length. It is important to notice that on the right side of Equation (3.17) there are three terms that determine the behavior of the charge in the channel: a linear term, which is important in the strong-inversion region, the first logarithm term, which is important in the subthreshold region, and the last logarithm term, which is important in the moderate-inversion region. Therefore, Equation (3.17) can represent the mobile carrier concentration in the channel for all bias conditions in a continuous and smooth manner, which is crucial for circuit simulation success.

The normalized drain current is obtained from the solution of the Poisson-carrier transport equation [23] and is represented by

$$i_{ds} = \left[\frac{q_m^2}{2} - 2q_m - q_H \ln \left(1 - \frac{q_m}{q_H} \right) \right] \Bigg|_{q_{m,s}}^{q_{m,d}},\tag{3.25}$$

where

$$q_H = \frac{1}{r_N} - q_{dep}.\tag{3.26}$$

The drain current normalization is given by

$$i_{ds} = \frac{-I_{ds}L}{\mu_m v_T^2 C_{ins}}.\tag{3.27}$$

Equation (3.25) has three terms that determine the behavior of the current under different bias conditions. The first term, quadratic, is important in the saturation and triode conditions, the second term, linear, is important in the triode and subthreshold conditions, and the last term, logarithmic, is important in the subthreshold and moderate-inversion conditions. This shows that Equation (3.25) represent the drain current characteristics in all regions of device operation—that is, subthreshold, linear, and saturation regions—in a continuous and explicit expression. Considering each region of operation, the proposed model can be reduced to simple expressions. In the subthreshold region, the drain current is approximately given by

$$I_{ds} \approx \frac{\mu}{L} v_T^2 C_{ins} \exp \left(\frac{V_g - V_{th}}{v_T} \right) \times \left(1 - \exp \frac{-V_{ds}}{v_T} \right),\tag{3.28}$$

where V_{th} is the threshold voltage of FinFETs [30]. Note that Equation (3.28) is independent of C_{ins} for undoped devices; thus, Equation (3.28) can be further simplified to

$$I_{ds} \approx A_{ch} \frac{\mu}{L} v_T q \frac{n_i^2}{N_{ch}} \exp \left(\frac{V_g - V_{fb}}{v_T} \right) \times \left(1 - \exp \frac{-V_{ds}}{v_T} \right).\tag{3.29}$$

With use of Equation (3.29) it is possible to conclude that to decrease the leakage in lightly doped FinFETs, A_{ch} must be scaled down. The drain currents in the linear and saturation regions are approximately given by

$$I_{ds} \approx \frac{\mu}{L} C_{ins} \left(V_g - V_{th} - V_{ds}/2 \right) V_{ds},\tag{3.30}$$

$$I_{ds} \approx \frac{\mu}{2L} C_{ins} \left(V_g - V_{th}\right)^2 . \qquad (3.31)$$

Equations (3.30) and (3.31) are the very well-known equations from the quadratic model for conventional long-channel CMOS MOSFETs.

From the results presented, it should be noted that only four different model parameters are needed for modeling FinFET devices: A_{ch}, N_{ch}, W, and C_{ins}. With use of these parameters, a FinFET with a simple cross-section, such as a DG FinFET, can be accurately modeled for different channel doping concentrations as shown in Figure 3.11. The model parameters used for DG FinFETs are as follows [23]:

$$A_{ch} = H_{fin} T_{fin}, \qquad (3.32)$$

$$W = 2H_{fin}, \qquad (3.33)$$

$$C_{ins} = \frac{\varepsilon_{ins}}{EOT} W, \qquad (3.34)$$

$$N_{ch}, \qquad (3.35)$$

where EOT is the equivalent oxide thickness.

A trapezoidal TG FinFET is a good example of a FinFET with a complex cross-section. Indeed, the industry transistors reported in [25, 26] have fin cross-sections

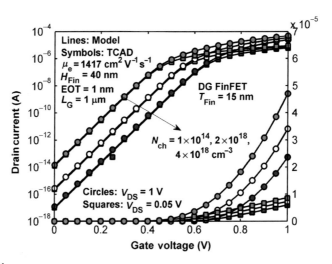

FIGURE 3.11

Drain current versus gate voltage of DG FinFETs at $V_{ds} = 0.05\,V$ (squares) and $V_{ds} = 1\,V$ (circles) obtained from the proposed model (lines) and numerical simulations (symbols) for three different channel dopings: $N_{ch} = 1 \times 10^{14}\,cm^{-3}$, $N_{ch} = 2 \times 10^{18}\,cm^{-3}$, and $N_{ch} = 4 \times 10^{18}\,cm^{-3}$. $\mu = 1470\,cm^2\,Vs^{-1}$, $L_G = 1\,\mu m$, $T_{ch} = 15\,nm$, $H_{ch} = 40\,nm$, EOT $= 1\,nm$, and a metal gate work function equal to 4.6 eV were used.

similar to trapezoidal shapes. The proposed model can be used to model these types of devices through the use of the following four model parameters:

$$A_{\mathrm{ch}} = H_{\mathrm{fin}} \frac{(T_{\mathrm{fin,top}} + T_{\mathrm{fin,base}})}{2}, \tag{3.36}$$

$$W = 2\sqrt{\frac{(T_{\mathrm{fin,base}} - T_{\mathrm{fin,top}})^2}{4} + H_{\mathrm{fin}}^2} + T_{\mathrm{fin,top}}, \tag{3.37}$$

$$C_{\mathrm{ins}} = \frac{\varepsilon_{\mathrm{ins}}}{\mathrm{EOT}} W, \tag{3.38}$$

$$N_{\mathrm{ch}}. \tag{3.39}$$

With use of these parameters, trapezoidal TG FinFETs can be accurately modeled as shown in Figure 3.12, where a trapezoidal TG FinFET has been doped at different doping concentrations as used in chips with multithreshold voltage levels.

In the case of a channel dimension variation, the model can accurately predict the trend of current changes as shown in Figure 3.13. Note that a $T_{\mathrm{fin,top}}$ variation is more important for the on current than a $T_{\mathrm{fin,base}}$ variation. In addition, the off current varies linearly as function of $T_{\mathrm{fin,top}}$ or $T_{\mathrm{fin,base}}$, as expected.

The proposed model accurately models experimental long-channel FinFETs without the use of fitting parameters as shown in Figures 3.14 and 3.15, which

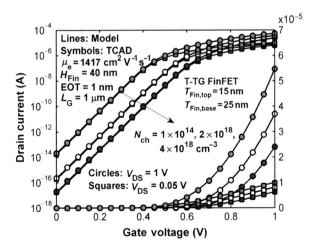

FIGURE 3.12

I_{d} versus V_{g} of trapezoidal TG FinFETs ($T_{\mathrm{fin,top}} = 15\,\mathrm{nm}$ and $T_{\mathrm{fin,base}} = 25\,\mathrm{nm}$) at $V_{\mathrm{ds}} = 0.05\,\mathrm{V}$ (squares) and $V_{\mathrm{ds}} = 1\,\mathrm{V}$ (circles) obtained from the proposed model (lines) and numerical simulations (symbols) for three different channel dopings: $N_{\mathrm{ch}} = 1 \times 10^{14}\,\mathrm{cm}^{-3}$, $N_{\mathrm{ch}} = 2 \times 10^{18}\,\mathrm{cm}^{-3}$, and $N_{\mathrm{ch}} = 4 \times 10^{18}\,\mathrm{cm}^{-3}$.

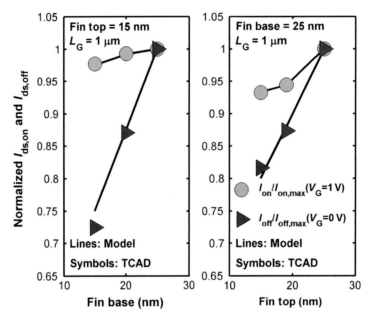

FIGURE 3.13

Normalized ($I_{ds}/I_{ds,max}$) on (circles) and off (triangles) drain currents.

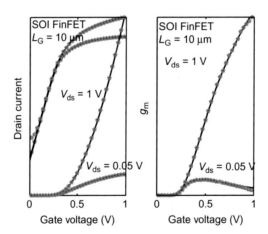

FIGURE 3.14

Drain current versus gate voltage (left) for a long-channel silicon-on-insulator (SOI) FinFET obtained from experimental data and from the proposed model using only the ideal unified long-channel core model. There is also good agreement for transconductance (right) between the model and experimental data. The symbols represent the experimental data and the lines represent the proposed model.

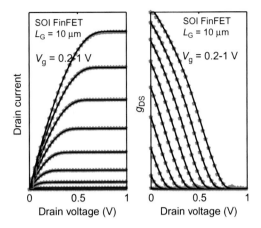

FIGURE 3.15

Drain current versus drain voltage (left) for a long-channel SOI FinFET obtained from experimental data and from the proposed model using only the ideal unified long-channel core model. There is also good agreement for the output conductance (right) between the model and experimental data. The symbols represent the experimental data and the lines represent the proposed model.

compare the proposed core compact model and the data from a fabricated long-channel FinFET. Note that only a mobility model was included additionally to the core model in producing the figures. The accuracy of the proposed model is tested by agreement between the data and the model for the drain current and its derivatives. This is a very important test because g_m and g_{ds} are crucial quantities in the ultimate performance of circuits constructed from these FinFETs.

CHAPTER 3 APPENDIX: EXPLICIT SURFACE POTENTIAL MODEL

The surface potential model used in BSIM-CMG was initially derived in an implicit form in Section 3.1, and relies on numerical calculations to be solved, such as the Newton-Raphson method [31, 32]. The Newton-Raphson method uses an iterative algorithm that start from an initial guess and then keeps refining it until the method arrives at a final solution. This refinement is done by approximating the function to be solved for the given guess by a linear system, which is then used to obtain the new guess for the solution [33]. Although the Newton-Raphson method could potentially result in a rapid convergence (e.g., quadratic [32, 33]), convergence is not guaranteed. Indeed, it could potentially fail by, for example, getting into an infinite loop. This is not practical for compact models, where convergence and speed are key requirements. However, by making sure that the initial guess is close enough to the final solution and that the solution is not singular—that is, the derivative of the function is nonzero at the solution—one can ensure that the Newton-Raphson method will result in convergence [32]. Therefore, in order to implement the model proposed in Section 3.1 in a circuit simulator, a good initial guess formula, the so-called continuous starting function (CSF) [24], for Equation (3.11) is presented in this section. In addition, the Newton-Raphson method can be replaced for a higher-order method such as the quartic modified iteration [32]. The main difference between the quartic modified and Newton-Raphson iterations is that first refines the initial guess by approximating the function to be solved using a high-order approximation, instead of the linear one used in the Newton-Raphson iteration. Therefore, quartic modified iteration could give faster convergence, which would reduce the number of iterations needed to solve Equation (3.11), making the proposed compact model fast.

3A.1 CONTINUOUS STARTING FUNCTION

An initial guess for Equation (3.11) was proposed in [3]. It was derived by considering the asymptotic behavior of β in Equation (3.11) under the two main regimes of operation: subthreshold regime and strong-inversion regime. Although the initial guess proposed in [3] works well for most conditions, it has a few important issues that must be solved. First, it is not a single continuous and smooth starting function because it was constructed by taking the minimum between two different functions. The lack of smoothness implies a higher number of iterations needed to arrive at a final solution. The second issue is that the initial guess does not work very well for devices with highly doped or wide channels, where β can be as low as 1×10^{-15}. This makes the model unstable depending on the device geometry and bias conditions. This a critical point, especially when a compact model is used for circuit-device optimization where the optimization algorithm could bring the evaluation of the model to extreme conditions. Because of the aforementioned issues, in this section a new CSF is proposed to solve Equation (3.11). In a similar manner as the initial guess proposed in [3], the new CSF will be obtained by considering the asymptotic behavior of Equation (3.11); however, the highly doped condition will be considered first to later apply the derived CSF to lightly doped channels.

In the case of a FinFET device with a highly doped channel, ψ_{pert} is large and β tends to 0, which makes $\cos\beta$ to tend to 1 and $\ln(\cos\beta)$ tend to 0. Using these assumptions and disregarding $\ln\beta$ term, which is close to being constant in the strong-inversion condition, we can obtain a β function for highly doped channel FinFETs in the strong-inversion region by solving Equation (3.11) for the β^2 term inside the square root:

$$\beta_{\text{si}} = e^{\frac{-\psi_{\text{pert}}}{2V_{\text{tm}}}} \frac{A_{g0}}{r} \sqrt{\left(\frac{F - F_{\text{th,si}}}{A_{g0}} + 1\right)^2 - 1}. \qquad (3A.1)$$

Here, A_{g0}, r, F, and $F_{th,si}$ are defined as follows:

$$A_{g0} = \frac{r\psi_{pert}}{V_{tm}}. \tag{3A.2}$$

$$r = \frac{2\epsilon_{si}t_{ox}}{\epsilon_{ox}T_{fin}}, \tag{3A.3}$$

$$F = \frac{V_g - V_{fb} - V_{ch} - \psi_{pert}}{2V_{tm}} + \ln\left(\frac{T_{fin}}{2}\sqrt{\frac{qn_i^2}{2\epsilon_{si}N_{ch}V_{tm}}}\right), \tag{3A.4}$$

$$F_{th,si} = (2r - 1)\frac{\psi_{pert}}{2V_{tm}}. \tag{3A.5}$$

Equation (3A.1) is valid only when $(F - F_{th,si})/A_{g0} > -1$. In the subthreshold region, β is even smaller than in the previous assumption, and β^2 in Equation (3.11) can be disregarded. Then $\ln\beta$ is no longer constant in the subthreshold region, and from it, a β function for the subthreshold region can be obtained as follows [3]:

$$\beta_{ST} = e^{F - F_{th}}e^{\frac{-\psi_{pert}}{2V_{tm}}}. \tag{3A.6}$$

Note that Equation (3A.6) is a good approximation for the subthreshold region of lightly and heavily doped channels as well. Equations (3A.1) and (3A.6) are good approximations for β; however, a single CSF has not been constructed yet. In order to obtain a single CSF, Equation (3A.1) can be modified to

$$\beta_{doped} = e^{\frac{-\psi_{pert}}{2V_{tm}}}\frac{A_{g0}}{r}\sqrt{\left(\frac{\ln\left(1 + e^{2(F - F_{th,si})}\right)}{2A_{g0}} + 1\right)^2 - 1}. \tag{3A.7}$$

The above expression can be evaluated in all bias conditions and it tends to Equation (3A.1) when $F > F_{th}$. In the case of $F \ll F_{th,si}$, Equation (3A.7) tends to an expression similar but not exactly equal to Equation (3A.6). In order to make Equation (3A.7) reduce to Equation (3A.6) for $F \ll F_{th,si}$, $F_{th,si}$ can be modified by equating Equations (3A.1) and (3A.6) at the subthreshold, giving

$$F_{th} = (2r - 1)\frac{\psi_{pert}}{2V_{tm}} + \ln\left(\frac{\sqrt{A_{g0}}}{(2r - 1)\frac{\psi_{pert}}{2V_{tm}}}\right). \tag{3A.8}$$

Figures 3A.1 and 3A.2 show β obtained from Equation (3.11) solved using Newton-Raphson iteration and from an initial guess expressed by Equation (3A.7) using Equation (3A.8). As expected, the proposed guess is close to the final solution in all bias regimes for highly doped channels. In order to extend Equation (3A.7) to

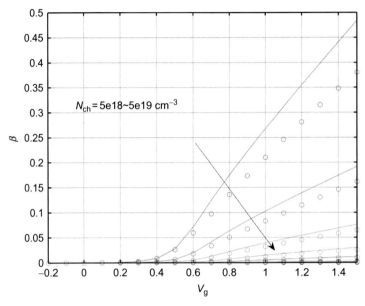

FIGURE 3A.1

β obtained from Equation (3.11) solved using Newton-Raphson iteration (symbols) and from an initial guess expressed by Equation (3A.7) (lines) using Equation (3A.8) on a linear scale, for different doping concentrations. Equation (3A.7) should be extended to cover lightly doped devices. $T_{\text{fin}} = 20\,\text{nm}$, $t_{\text{ox}} = 1\,\text{nm}$, and a gate work function equal to 4.4 eV were used for the simulations.

devices with lightly doped channels, only two major changes are needed. First, it should be noticed that as the amount of channel doping decreases, A_{g0} tends to 0, and this would give invalid results. Therefore, analyzing the asymptotic behavior of Equation (3.11), we can see that as doping decreases, a valid value for A_{g0} should be 1, and thus A_{g0} can becomes

$$A_g = \frac{\frac{r\psi_{\text{pert}}}{V_{\text{tm}}}}{1 - e^{\frac{-r\psi_{\text{pert}}}{V_{\text{tm}}}}}. \tag{3A.9}$$

The second change that must be done is to limit the value of β obtained from Equation (3A.7) to $\pi/2$. This limit is obtained by analyzing the behavior of Equation (3.11) in strong inversion for lightly doped devices as also explained in [3]. Therefore, the final CSF proposed here is given as follows:

$$\beta_0 = \frac{1}{\frac{1}{\beta_{\text{doped}}} + \frac{2}{\pi}}, \tag{3A.10}$$

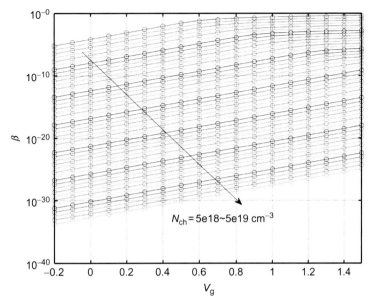

FIGURE 3A.2

β obtained from Equation (3.11) solved using Newton-Raphson iteration (symbols) and from an initial guess expressed by Equation (3A.7) (lines) using Equation (3A.8) on a logarithmic scale, for different doping concentrations. Equation (3A.7) should be extended to cover lightly doped devices. $T_{fin} = 20$ nm, $t_{ox} = 1$ nm, and a gate work function equal to 4.4 eV were used for the simulations.

where β_0 is obtained from Equation (3A.7) using F_{th} and A_g from Equations (3A.8) and (3A.9), respectively. Figures 3A.3 and 3A.4 show β obtained from Equation (3.11) solved using Newton-Raphson iteration and from an initial guess expressed by Equation (3A.10). As expected, Equation (3A.10) is valid for devices with lightly or heavily doped channels.

3A.2 QUARTIC MODIFIED ITERATION: IMPLEMENTATION AND EVALUATION

In order to complete the explicit surface potential model, a quartic modified iteration has to be derived. The quartic modified iteration updates the initial guess expressed by Equation (3A.10) using a high-order correction:

$$\beta_1 = \beta_0 - \frac{f_0}{f_1}\left(1 + \frac{f_0 f_2}{2f_1^2} + \frac{f_0^2\left(3f_2^2 - f_1 f_3\right)}{6f_1^4}\right), \tag{3A.11}$$

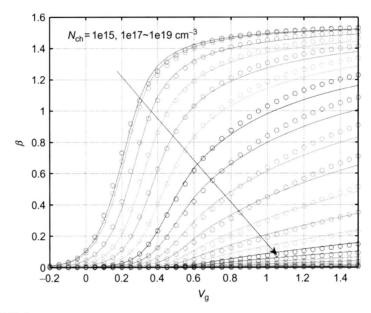

FIGURE 3A.3

β obtained from Equation (3.11) solved using Newton-Raphson iteration (symbols) and from the proposed CSF expressed by Equation (3A.10) (lines) on a linear scale, for different doping concentrations. Equation (3A.10) is valid for devices with lightly or heavily doped channels. Note that for lightly doped devices β has a similar behavior, which changes as the amount of doping is increased. $T_{\text{fin}} = 20\,\text{nm}$, $t_{\text{ox}} = 1\,\text{nm}$, and a gate work function equal to 4.4 eV were used for the simulations.

where f_n is the nth derivative of Equation (3.11) with respect to β—that is, $f_n = \frac{\partial^n f}{\partial \beta^n}|_{\beta=\beta_0}$. Using this algorithm, one can calculate an explicit surface potential model using the following steps:

1. Calculate the CSF in the following order:

$$T_{\text{a}} = \frac{\psi_{\text{pert}}}{V_{\text{tm}}^2};\tag{3A.12a}$$

$$T_{\text{b}} = e^{\frac{\psi_{\text{pert}}}{V_{\text{tm}}}};\tag{3A.12b}$$

$$T_{\text{c}} = 2V_{\text{tm}};\tag{3A.12c}$$

$$r = \frac{2\epsilon_{\text{si}}t_{\text{ox}}}{\epsilon_{\text{ox}}T_{\text{fin}}};\tag{3A.12d}$$

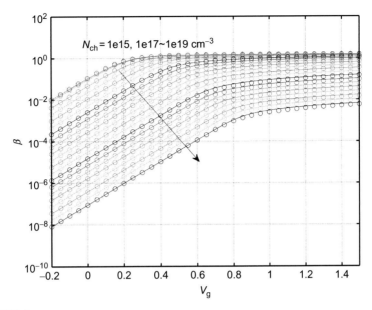

FIGURE 3A.4

β obtained from Equation (3.11) solved using Newton-Raphson iteration (symbols) and from the proposed CSF expressed by Equation (3A.10) (lines) on a logarithmic scale, for different doping concentrations. Equation (3A.10) is valid for devices with lightly or heavily doped channels. Note that for lightly doped devices β has a similar behavior, which changes as the amount of doping is increased. $T_{fin} = 20\,nm$, $t_{ox} = 1\,nm$, and a gate work function equal to $4.4\,eV$ were used for the simulations.

$$A_{\mathrm{g}} = \dfrac{\dfrac{r\psi_{\mathrm{pert}}}{V_{\mathrm{tm}}}}{1 - \mathrm{e}^{\dfrac{-r\psi_{\mathrm{pert}}}{V_{\mathrm{tm}}}}}\,; \tag{3A.12e}$$

$$F_{\mathrm{th}} = (2r - 1)\dfrac{\psi_{\mathrm{pert}}}{2V_{\mathrm{tm}}} + \ln\left(\dfrac{\sqrt{A_{\mathrm{g}}}}{(2r-1)\dfrac{\psi_{\mathrm{pert}}}{2V_{\mathrm{tm}}}}\right)\,; \tag{3A.12f}$$

$$\beta_{\mathrm{doped}} = \mathrm{e}^{\dfrac{-\psi_{\mathrm{pert}}}{2V_{\mathrm{tm}}}}\dfrac{A_{\mathrm{g}}}{r}\sqrt{\left(\dfrac{\ln\left(1 + \mathrm{e}^{2(F - F_{\mathrm{th}})}\right)}{2A_{\mathrm{g}}} + 1\right)^{2} - 1}\,; \tag{3A.12g}$$

$$\beta_0 = \cfrac{1}{\cfrac{1}{\beta_{\text{doped}}} + \cfrac{2}{\pi}}. \tag{3A.12h}$$

2. Compute

$$\tan g0 = \tan \beta_0, \tag{3A.13a}$$

$$\cos g0 = \cos \beta_0, \tag{3A.13b}$$

$$\sec g0 = \cos g0^{-1}, \tag{3A.13c}$$

$$\sec g0sq = \sec g0^2, \tag{3A.13d}$$

$$\ln g0 = \ln \beta_0. \tag{3A.13e}$$

3. Compute

$$T_0 = 1 + \beta_0 \tan g0, \tag{3A.14a}$$

$$T_1 = \beta_0^2 (T_b \sec g0sq - 1) + T_{\text{abb}} (\psi_{\text{pert}} - T_c \ln(\cos g0)), \tag{3A.14b}$$

$$T_2 = \sqrt{T_1}, \tag{3A.14c}$$

$$T_3 = -2\beta + T_{\text{abb}} T_c \tan g0 + 2T_b \beta \sec g0sq T_0, \tag{3A.14d}$$

$$T_4 = -2 + 2T_b \beta_0^2 \sec g0sq^2 + \sec g0sq(2T_b + T_a T_c + 8T_b \beta \tan g0 + 4T_b \beta_0^2 \tan g0^2), \tag{3A.14e}$$

$$T_5 = 2 \tan g0 T_4 + 4(3T_b \beta \sec g0sq^2 T_0 + \tan g0 + 2T_b \sec g0sq T_0 \tan g0). \tag{3A.14f}$$

4. Compute the following derivatives:

$$f_0 = \ln g0 - \ln(\cos g0) + rT_2 - F \tag{3A.15a}$$

$$f_1 = \beta^{-1} + \tan g0 + \frac{rT_3}{2T_2} \tag{3A.15b}$$

$$f_2 = -\beta_0^{-2} + \sec g0sq - \frac{rT_3^2}{4T_2^3} + \frac{rT_4}{2T_2} \tag{3A.15c}$$

$$f_0 = 2\beta_0^{-3} + 2 \sec g0sq \tan g0 + \frac{3rT_3}{4T_2^3} \left(\frac{T_3^2}{2T_2^2} - T_4 \right) + \frac{rT_5}{2T_2}. \tag{3A.15d}$$

5. Update β_0 to β_1 using a quartic modified iteration:

$$\beta_1 = \beta_0 - \frac{f_0}{f_1}\left(1 + \frac{f_0 f_2}{2f_1^2} + \frac{f_0^2\left(3f_2^2 - f_1 f_3\right)}{6f_1^4}\right). \qquad (3A.16a)$$

Steps 2-5 can be performed more than once. The number of iterations can be obtained by doing an extensive analysis of the proposed model accuracy under different device and bias conditions. Figure 3A.5 shows the error of β obtained from the proposed explicit surface potential model with one and two iterations with respect to β obtained using the Newton-Raphson method under all different combinations (more than 52,500 simulations) of doping concentration (1×10^{15} to $1 \times 10^{19}\,\mathrm{cm}^{-3}$), channel width (1-30 nm), dielectric thickness (0.5-5 nm), gate voltage (-0.2 to 1.5 V), and temperature (-100 to $100\,^\circ\mathrm{C}$). With one Householder iteration, the RMS error is 0.28%, with a peak error of 2.47%. With two Householder iterations, the RMS error is $1.58 \times 10^{-6}\%$, with a peak error of $3.31 \times 10^{-5}\%$. Therefore, only two iterations are needed to obtain an accurate solution of β as shown in Figure 3A.6.

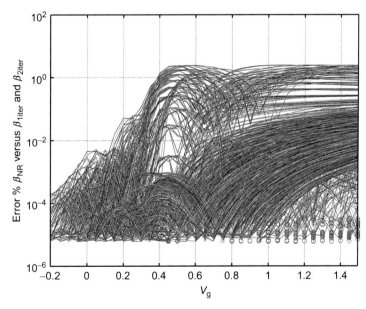

FIGURE 3A.5

Error of β obtained from the proposed explicit surface potential model with one (lines) and two (symbols) iterations with respect to β obtained using the Newton-Raphson method under all different combinations of doping concentration (1×10^{15} to $1 \times 10^{19}\,\mathrm{cm}^{-3}$), channel width (1-30 nm), dielectric thickness (0.5-5 nm), gate voltage (-0.2 to 1.5 V), and temperature (-100 to $100\,^\circ\mathrm{C}$).

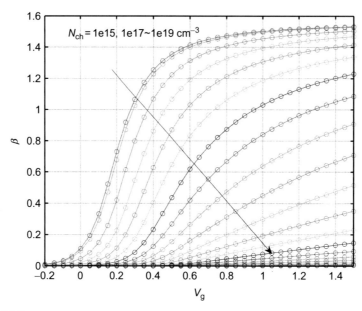

FIGURE 3A.6

β obtained from Equation (3.11) solved using Newton-Raphson iteration (symbols) and from the proposed explicit surface potential model using two Householder iterations (lines) on a linear scale, for different doping concentrations. $T_{fin} = 20$ nm, $t_{ox} = 1$ nm, and a gate work function equal to 4.4 eV were used for the simulations.

REFERENCES

[1] A. Ortiz-Conde, F. Garcia-Sanchez, J. Muci, S. Malobabic, J. Liou, A review of core compact models for undoped double-gate SOI MOSFETs, IEEE Trans. Electron Devices 54 (1) (2007) 131–140.

[2] H. Pao, C. Sah, Effects of diffusion current on characteristics of metal-oxide (insulator)-semiconductor transistors, Solid State Electron. 9 (10) (1966) 927–937.

[3] M. Dunga, Nanoscale CMOS Modeling, University of California, Berkeley, 2008.

[4] Y. Taur, An analytical solution to a double-gate MOSFET with undoped body, IEEE Electron Device Lett. 21 (5) (2000) 245–247.

[5] Y. Chen, J. Luo, A comparative study of double-gate and surrounding-gate MOSFETs in strong inversion and accumulation using an analytical model, Integration 1 (2) (2001) 6.

[6] M. Dunga, C. Lin, X. Xi, D. Lu, A. Niknejad, C. Hu, Modeling advanced FET technology in a compact model, IEEE Trans. Electron Devices 53 (9) (2006) 1971–1978.

[7] O. Moldovan, A. Cerdeira, D. Jimenez, J. Raskin, V. Kilchytska, D. Flandre, N. Collaert, B. Iñiguez, Compact model for highly-doped double-gate SOI MOSFETs targeting baseband analog applications, Solid State Electron. 51 (5) (2007) 655–661.

[8] F. Liu, L. Zhang, J. Zhang, J. He, M. Chan, Effects of body doping on threshold voltage and channel potential of symmetric DG MOSFETs with continuous solution from accumulation to strong-inversion regions, Semicond. Sci. Technol. 24 (2009) 085005.

[9] F. Liu, J. He, L. Zhang, J. Zhang, J. Hu, C. Ma, M. Chan, A charge-based model for long-channel cylindrical surrounding-gate MOSFETs from intrinsic channel to heavily doped body, IEEE Trans. Electron Devices 55 (8) (2008) 2187–2194.

[10] C.-H. Lin, R. Kambhampati, R. Miller, T. Hook, A. Bryant, W. Haensch, P. Oldiges, I. Lauer, T. Yamashita, V. Basker, et al., Channel doping impact on FinFETs for 22 nm and beyond, in: IEEE 2012 Symposium on VLSI Technology (VLSIT), 2012, pp. 15–16.

[11] B. Yu, J. Song, Y. Yuan, W. Lu, Y. Taur, A unified analytic drain current model for multiple-gate MOSFETs, IEEE Trans. Electron Devices 55 (8) (2008) 2157–2163.

[12] A. Tsormpatzoglou, C. Dimitriadis, R. Clerc, G. Pananakakis, G. Ghibaudo, Semianalytical modeling of short-channel effects in lightly doped silicon trigate MOSFETs, IEEE Trans. Electron Devices 55 (10) (2008) 2623–2631.

[13] E. Moreno, J. Roldán, F. Ruiz, D. Barrera, A. Godoy, F. Gámiz, An analytical model for square GAA MOSFETs including quantum effects, Solid State Electron. 54 (11) (2010) 1463–1469.

[14] E. Moreno Perez, J. Roldan Aranda, F. Garcia Ruiz, D. Barrera Rosillo, M. Ibanez Perez, A. Godoy, F. Gamiz, An inversion-charge analytical model for square gate-all-around MOSFETs, IEEE Trans. Electron Devices 58 (9) (2011) 2854–2861.

[15] J. Park, J. Colinge, C. Diaz, Pi-gate SOI MOSFET, IEEE Electron Device Lett. 22 (8) (2001) 405–406.

[16] Y. Taur, X. Liang, W. Wang, H. Lu, A continuous, analytic drain-current model for DG MOSFETs, IEEE Electron Device Lett. 25 (2) (2004) 107–109.

[17] J. Sallese, F. Krummenacher, F. Prégaldiny, C. Lallement, A. Roy, C. Enz, A design oriented charge-based current model for symmetric DG MOSFET and its correlation with the EKV formalism, Solid State Electron. 49 (3) (2005) 485–489.

[18] A. Ortiz-Conde, F. García Sánchez, J. Muci, Rigorous analytic solution for the drain current of undoped symmetric dual-gate MOSFETs, Solid State Electron. 49 (4) (2005) 640–647.

[19] G. Smit, A. Scholten, G. Curatola, R. van Langevelde, G. Gildenblat, D. Klaassen, PSP-based scalable compact FinFET model, Proc. NSTI Nanotech. 3 (2007) 520–525.

[20] H. Abd El Hamid, J. Guitart, V. Kilchytska, D. Flandre, B. Iniguez, A 3-D analytical physically based model for the subthreshold swing in undoped trigate FinFETs, IEEE Trans. Electron Devices 54 (9) (2007) 2487–2496.

[21] D. Jiménez, B. Iniguez, J. Sune, L. Marsal, J. Pallares, J. Roig, D. Flores, Continuous analytic IV model for surrounding-gate MOSFETs, IEEE Electron Device Lett. 25 (8) (2004) 571–573.

[22] S. Venugopalan, D. Lu, Y. Kawakami, P. Lee, A. Niknejad, C. Hu, BSIM-CG: a compact model of cylindrical/surround gate MOSFET for circuit simulations, Solid State Electron. 67 (1) (2012) 79–89.

[23] J.P. Duarte, N. Paydavosi, S. Venugopalan, A. Sachid, C. Hu, Unified FinFET compact model: modelling trapezoidal triple-gate FinFETs, in: SISPAD, 2013.

[24] B. Yu, H. Lu, M. Liu, Y. Taur, Explicit continuous models for double-gate and surrounding-gate MOSFETs, IEEE Trans. Electron Devices 54 (10) (2007) 2715–2722.

[25] C.-Y. Chang, T.-L. Lee, C. Wann, L.-S. Lai, H.-M. Chen, C.-C. Yeh, C.-S. Chang, C.-C. Ho, J.-C. Sheu, T.-M. Kwok, et al., A 25-nm gate-length FinFET transistor module for 32 nm node, in: 2009 IEEE International Electron Devices Meeting (IEDM), 2009, pp. 1–4.

[26] C. Auth, C. Allen, A. Blattner, D. Bergstrom, M. Brazier, M. Bost, M. Buehler, V. Chikarmane, T. Ghani, T. Glassman, et al., A 22 nm high performance and low-power CMOS technology featuring fully-depleted tri-gate transistors, self-aligned contacts and high density MIM capacitors, in: IEEE 2012 Symposium on VLSI Technology (VLSIT), 2012, pp. 131–132.

[27] N. Paydavosi, S. Venugopalan, Y.S. Chauhan, J. Duarte, S. Jandhyala, A. Niknejad, C. Hu, BSIM-SPICE models enable FinFET and UTB IC designs, IEEE Access 1 (2013) 201.

[28] N. Chevillon, J.-M. Sallese, C. Lallement, F. Prégaldiny, M. Madec, J. Sedlmeir, J. Aghassi, Generalization of the concept of equivalent thickness and capacitance to multigate MOSFETs modeling, IEEE Trans. Electron Devices 59 (1) (2012) 60–71.

[29] J.P. Duarte, S.-J. Choi, D.-I. Moon, J.-H. Ahn, J.-Y. Kim, S. Kim, Y.-K. Choi, A universal core model for multiple-gate field-effect transistors. Part I: charge model, IEEE Trans. Electron Devices 60 (2) (2013) 840–847.

[30] J.P. Duarte, S.-J. Choi, D.-I. Moon, J.-H. Ahn, J.-Y. Kim, S. Kim, Y.-K. Choi, A universal core model for multiple-gate field-effect transistors. Part II: drain current model, IEEE Trans. Electron Devices 60 (2) (2013) 848–855.

[31] W.H. Press, B.P. Flannery, S.A. Teukolsky, W. Vetterling, B. Flannery, Numerical recipes: the art of scientific computing (FORTRAN), 1989.

[32] P. Sebah, X. Gourdon, Newtons method and high order iterations, Technical report, 2001. Available from: http://numbers.computation.free.fr/Constants/Algorithms/newton.html.

[33] J. Roychowdhury, Numerical Simulation and Modelling of Electronic and Biochemical Systems, Now Publishers, Inc., Hanover, MA, 2009.

Channel current and real device effects

4

CHAPTER OUTLINE

4.1 Introduction...99
4.2 Threshold voltage roll-off.. 100
4.3 Subthreshold slope degradation.. 106
4.4 Quantum mechanical v_{th} correction 107
4.5 Vertical-field mobility degradation 109
4.6 Drain saturation voltage, v_{dsat} 109
    4.6.1 Extrinsic case (RDSMOD = 1 and 2) 110
    4.6.2 Intrinsic case (RDSMOD = 0)...................................... 111
4.7 Velocity saturation model 114
4.8 Quantum mechanical effects 115
    4.8.1 Effective width model 118
    4.8.2 Effective oxide thickness/effective capacitance 118
    4.8.3 Charge centroid calculation for accumulation 119
4.9 Lateral nonuniform doping model.. 119
4.10 Body effect model for a bulk FinFET (BULKMOD = 1) 119
4.11 Output resistance model.. 120
    4.11.1 Channel-length modulation 121
    4.11.2 Drain-induced barrier lowering................................... 122
4.12 Channel current 123
References... 124

4.1 INTRODUCTION

This chapter discusses the channel current modeling in the BSIM-CMG model. Modeling of the core current valid for a wide-long double-gate MOSFET was discussed in Chapter 3. Some of the important effects affecting channel current in short-channel/medium-channel-length devices are described here. The short-channel effects in FinFETs are the same as those in bulk MOSFETs but are less severe owing to better electrostatic control on the channel from multiple gates in a FinFET. Some of the dominant effects are

- threshold voltage (V_{th}) roll-off,
- subthreshold slope (SS) (n) degradation,
- mobility degradation due to the vertical electric field,
- carrier velocity saturation,
- the series resistance effect (see Chapter 7),
- drain-induced barrier lowering (DIBL),
- channel-length modulation (CLM), and
- quantum mechanical effects.

4.2 THRESHOLD VOLTAGE ROLL-OFF

In a short-channel device, the drain starts affecting the potential barrier seen by carriers while entering from the source side as the source is in close proximity to the drain. Although there are numerous models for modeling of threshold voltage roll-off in double-gate devices, BSIM-CMG uses a physical but simple model such that it is computationally efficient and implementable.

Short-channel effects originate from the influence of a two-dimensional field in the channel. The two-dimensional Poisson equation inside the body in the subthreshold regime, where inversion carriers can be disregarded compared with bulk charge, can be written as

$$\frac{d^2\psi(x,y)}{dx^2} + \frac{d^2\psi(x,y)}{dy^2} = \frac{qN_A}{\epsilon_{Si}}, \tag{4.1}$$

where N_A is the channel doping concentration. Here the x-axis is perpendicular and the y-axis is parallel to the channel. $\psi(x,y)$ is the electrostatic potential at any point (x,y) in the channel.

Characteristic length

The characteristic field penetration length, also known as the scale length, is an important parameter which defines the amount of the short-channel effect in the transistor and captures the variation in the amount of the drain field penetrating into the silicon body. It is a function of physical parameters such as T_{Si} and T_{ox} [1, 2].

To develop the characteristic length model, a parabolic potential profile in the channel perpendicular to the silicon-insulator interface (x-direction in Figure 4.1) is assumed as follows:

$$\psi(x,y) = C_0(y) + C_1(y)x + C_2(y)x^2. \tag{4.2}$$

C_0, C_1, and C_2 are independent of x but are functions of y. These can be obtained by applying three boundary conditions along the silicon body:

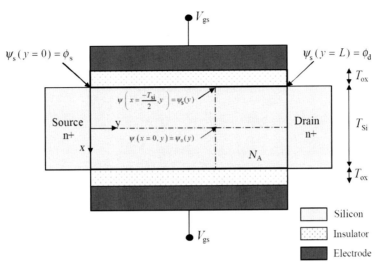

FIGURE 4.1

A symmetric common-gate double-gate FET.

$$\psi\left(x = \frac{T_{Si}}{2}, y\right) = \psi_s(y),\tag{4.3}$$

$$\left.\frac{d\psi(x, y)}{dx}\right|_{x=0} = 0,\tag{4.4}$$

$$\left.\frac{d\psi(x, y)}{dx}\right|_{x=\frac{T_{Si}}{2}} = \frac{V_{gs} - V_{fb} - \psi_s}{T_{ox}}\frac{\epsilon_{ox}}{\epsilon_{Si}},\tag{4.5}$$

where ψ_s is the surface potential. Using the parabolic profile in the vertical direction and applying the boundary condition of $\frac{d\psi}{dx} = 0$ for $x = \frac{T_{Si}}{2}$, we obtain $\psi(x, y)$ as

$$\psi(x, y) = \psi_s(y) + \frac{\epsilon_{ox}}{\epsilon_{Si}}\frac{\psi_s(y) - (V_{gs} - V_{fb})}{T_{ox}}x - \frac{\epsilon_{ox}}{\epsilon_{Si}}\frac{\psi_s(y) - (V_{gs} - V_{fb})}{T_{ox}T_{Si}}x^2.\tag{4.6}$$

If ψ_c is the potential distribution along the channel at the center, the relation between ψ_s and ψ_c will be obtained by substituting $x = \frac{T_{Si}}{2}$:

$$\psi_s(y) = \frac{1}{1 + \frac{\epsilon_{ox}}{4\epsilon_{Si}}\frac{T_{Si}}{T_{ox}}}\left(\psi_c(y) + \frac{\epsilon_{ox}}{4\epsilon_{Si}}\frac{T_{Si}}{T_{ox}}(V_{gs} - V_{fb})\right).\tag{4.7}$$

Now $\psi(x, y)$ can be expressed as

$$\psi(x, y) = \left(1 + \frac{\epsilon_{ox}}{\epsilon_{Si}}\frac{x}{T_{ox}} - \frac{\epsilon_{ox}}{\epsilon_{Si}}\frac{x^2}{T_{ox}T_{Si}}\right) \cdot \frac{\psi_c(y) + \frac{\epsilon_{ox}}{4\epsilon_{Si}}\frac{T_{Si}}{T_{ox}}(V_{gs} - V_{fb})}{1 + \frac{\epsilon_{ox}}{4\epsilon_{Si}}\frac{T_{Si}}{T_{ox}}}$$

$$-\frac{\epsilon_{ox}}{\epsilon_{Si}}\frac{x}{T_{ox}}(V_{gs}-V_{fb})+\frac{\epsilon_{ox}}{\epsilon_{Si}}\frac{x^2}{T_{ox}T_{Si}}(V_{gs}-V_{fb}). \tag{4.8}$$

So, by using the equations above, we get

$$\frac{d^2\psi_c(y)}{dx^2}+\frac{V_{gs}-V_{fb}-\psi_c(y)}{\lambda^2}=\frac{qN_A}{\epsilon_{Si}}, \tag{4.9}$$

$$\lambda=\sqrt{\frac{\epsilon_{ox}}{\epsilon_{Si}}\left(1+\frac{\epsilon_{ox}T_{Si}}{4\epsilon_{Si}T_{ox}}\right)T_{Si}T_{ox}}. \tag{4.10}$$

This expression for the characteristic length (λ) is derived for a double-gate FET. Increasing the number of gates in a transistor improves the electrostatic control, thereby decreasing the penetration of the drain field. Thus, λ is different for different multigate architectures. To develop λ of a triple-gate structure, we need to include the gate control on the channel from the top and the sidewalls. For this, a three-dimensional Poisson equation with appropriate boundary conditions needs to be solved. Similarly, we need to consider the control of all four gates in quadruple-gate FETs. Solving Poisson's equation for triple-gate or quadruple-gate FETs is complex and cannot be done in a compact model. The alternative approach is to use the existing λ of a double-gate FET and a single gate to develop λ for other multiple-gate FETs. The compact form of unified λ developed for all multigate devices and used in BSIM-CMG is as follows [3, 4]:

$$\lambda_c=\frac{1}{\sqrt{\left(\frac{1}{\lambda}\right)^2+\left(\frac{\Lambda}{\lambda_{H_{fin}}}\right)^2}}, \tag{4.11}$$

where

$$\lambda_{H_{fin}}=\sqrt{\frac{\epsilon_{Si}}{4\epsilon_{ox}}\left(1+\frac{\epsilon_{ox}T_{Si}}{2\epsilon_{Si}T_{ox}}\right)H_{fin}T_{ox}}, \tag{4.12}$$

H_{fin} is the height of fin and

$$\Lambda=0 \text{ for a double-gate FET (GEOMOD}=0), \tag{4.13}$$

$$\Lambda=\frac{1}{2} \text{ for a triple-gate FET (GEOMOD}=1), \tag{4.14}$$

$$\Lambda=1 \text{ for a surround-gate FET (GEOMOD}=2,3). \tag{4.15}$$

See Appendix for description of different model parameters and switches (e.g., GEOMOD used above).

Channel potential

The electrostatic potential profile along the channel is developed using the two-dimensional Poisson equation (4.1) and taking a small rectangular box of width δy and height $\frac{T_{Si}}{2}$ in the channel. After the application of Gauss's law on the rectangular box at position y along the channel, we get the following:

$$\frac{T_{\text{Si}}}{2} \cdot \left[E_y(y) - E_y(y + \delta y) \right] + \delta y \left[E_x \left(\frac{-T_{\text{Si}}}{2} \right) - E_x(0) \right] = q \frac{T_{\text{Si}} \delta y N_A}{2 \epsilon_{\text{Si}}}. \qquad (4.16)$$

To simplify Equation (4.16), we divide both side by δy, and the result can be written as follows:

$$\frac{T_{\text{Si}}}{2} \cdot \frac{\left[E_y(y) - E_y(y + \delta y) \right]}{\delta y} + \left[E_x \left(\frac{-T_{\text{Si}}}{2} \right) - E_x(0) \right] = q \frac{T_{\text{Si}} N_A}{2 \epsilon_{\text{Si}}}. \qquad (4.17)$$

With the use of the relationship $E_y = -\frac{\partial \psi_s(y)}{\partial y}$, and by replacing $E_x \left(\frac{-T_{\text{Si}}}{2} \right)$ and $E_x(0)$ with their values from Equations (4.4) and (4.5), we get following relation from Equation (4.17):

$$\lambda^2 \cdot \frac{\partial^2 \psi_s(y)}{\partial y^2} = \psi_s(y) - \left(V_{\text{gs}} - V_{\text{fb}} - \frac{q N_A T_{\text{Si}}}{2 C_{\text{ox}}} \right), \qquad (4.18)$$

where λ is the characteristic length. The surface potentials at the source and drain ends are $\psi_{s0} = V_s + V_{\text{bi}}$ and $\psi_{s0} = V_{\text{ds}} + V_{\text{bi}}$, respectively. Solving Equation (4.18), we obtain $\psi_s(y)$, and using Equation 4.7, we can get $\psi_c(y)$ as follows [5]:

$$\psi_c(y) = V_{\text{SL}} + (V_{\text{bi}} - V_{\text{SL}}) \frac{\sinh \left(\frac{L-y}{\lambda} \right)}{\sinh \left(\frac{L}{\lambda} \right)} + (V_{\text{bi}} + V_{\text{ds}} - V_{\text{SL}}) \frac{\sinh \left(\frac{y}{\lambda} \right)}{\sinh \left(\frac{L}{\lambda} \right)}. \qquad (4.19)$$

V_{bi} is the built-in potential at the source end, and V_{SL} is equivalent to the center potential for a long-channel transistor:

$$V_{\text{SL}} = V_{\text{gs}} - V_{\text{fb}} - \frac{q N_A}{\epsilon_{\text{Si}}} \lambda^2. \qquad (4.20)$$

As $\sinh(0) = 0$, it is apparent that $\psi_c(y)$ reduces to V_{bi} for $y = 0$ and to V_{ds} for $y = L$. Note the similarity between the center potential given by Equation (4.19) and the surface potential obtained from similar two-dimensional analysis in bulk MOSFETs [5–7]. For a long-channel transistor ($L \gg \lambda$) in the middle of the channel, the terms in the ratios in Equation (4.19) are both $e^{-L/2\lambda}$, which is very small; hence, ψ_c approaches V_{SL} as expected. For a short-channel transistor (L is not large compared with λ), ψ_c will be larger than V_{SL}.

The threshold voltage (V_{th}) is the value of the gate voltage above which the surface potential is approximately pinned in strong inversion [5, 6]. In all BSIM models, this is denoted by ψ_{st} (see Equation 4.22). The potential Equation (4.19) is a minimum at approximately $x = L/2$, and the inversion charge density is also a minimum at this point [5, 7].

The difference in the threshold voltage (ΔV_T) between a short-channel transistor and a long-channel transistor is equivalent to the difference in ψ_c in these transistors at the onset of strong inversion such that the value of $\psi_c(x = L/2)$ from Equation (4.19) is equal to its long-channel value of ψ_{st}:

$$\Delta V_T = -\frac{2(V_{bi} - V_{SL}) + V_{ds}}{2\cosh\left(\frac{L}{2\lambda}\right) - 2}. \tag{4.21}$$

This expression has two important effects modeled through a single equation: first, the length dependence of V_{th} and second, the drain bias effect on V_{th}. To provide more flexibility in the fitting of experimental data, BSIM-CMG models these through separate ΔV_T terms along with different parameters.

Threshold voltage roll-off

Decrease of the threshold voltage with decrease in the channel length is called threshold voltage roll-off. For V_{th} roll-off modeling at small V_{ds}, we can put $V_{ds} = 0$ in Equation (4.21). Also, the center potential at the source end (V_{SL}) can be taken as the surface potential at the source side in the subthreshold region (ψ_{st}). Thus,

$$\psi_{st} = 0.4 + \text{PHIN} + \Phi_B. \tag{4.22}$$

The V_{th} roll-off in BSIM-CMG is modeled as

$$\Delta V_{th,SCE} = -\frac{\frac{1}{2} \cdot \text{DVT0}}{\cosh\left(\text{DVT1} \cdot \frac{L_{eff}}{\lambda}\right) - 1} \cdot (V_{bi} - \psi_{st}), \tag{4.23}$$

where two parameters, DVT0 and DVT1, have been added for flexibility in parameter extraction (see Appendix for description of parameters). An important point to keep in mind during implementation is the numerical robustness. The $\cosh(x)$ used above is a smooth function and is symmetric around $x = 0$. For large x, it goes to 1 and can cause a divide-by-zero problem. To avoid this issue, BSIM-CMG uses the following approximation for $x > 40$:

$$\frac{\frac{1}{2}}{\cosh(x) - 1} = \frac{\frac{1}{2}}{\frac{e^x + e^{-x}}{2} - 1} \approx -\frac{\frac{1}{2}}{\frac{e^x}{2}} = e^{-x}. \tag{4.24}$$

DIBL effect on the threshold voltage

In short-channel devices, the drain voltage starts affecting the peak of the barrier height seen by the carriers as shown in Chapter 1. This is called DIBL. The decrease in barrier height results in a decrease in the threshold voltage. As mentioned earlier, Equation (4.21) also captures the drain bias effect on the threshold voltage. Another impact of DIBL is seen on the I_{ds}-V_{ds} characteristics, where it causes a finite slope in I_{ds}-V_{ds} and increases output conductance. This is discussed in Section 4.11.

To model DIBL, Equation (4.21) is modified as follows with two extra parameters for more flexibility in parameter extraction:

$$\Delta V_{th,DIBL} = -\frac{0.5 \cdot \text{ETA0}_a}{\cosh\left(\text{DSUB} \cdot \frac{L_{eff}}{\lambda}\right) - 1} \cdot V_{dsx} + \text{DVTP0} \cdot V_{dsx}^{\text{DVTP1}}. \tag{4.25}$$

The second term on the right-hand side of equation above is added to model long-channel DIBL, which is also called drain-induced threshold shift [5, 8].

Reverse short-channel effect

To reduce short-channel effects in a given technology, halo implants are used near the source and the drain. Although FinFETs are considered lightly doped devices, there could be intentional halo doping in the channel for threshold voltage control or unintentional doping caused by punch-through implants. The halo doping results in an increase in threshold voltage with decreasing channel length as the average doping in the channel increases. This is called a reverse short-channel effect.

The reverse short-channel effect model is taken from BSIM4 and is given by

$$\Delta V_{\text{th,RSCE}} = \text{K1RSCE} \cdot \left[\sqrt{1 + \frac{\text{LPE0}}{L_{\text{eff}}}} - 1 \right] \cdot \sqrt{\psi_{\text{st}}}. \tag{4.26}$$

Finally, all effects are combined in a single threshold voltage shift as follows:

$$\Delta V_{\text{th,all}} = \Delta V_{\text{th,SCE}} + \Delta V_{\text{th,DIBL}} + \Delta V_{\text{th,RSCE}} + \Delta V_{\text{th,temp}}, \tag{4.27}$$

$$V_{\text{gsfb}} = V_{\text{gs}} - \Delta\Phi - \Delta V_{\text{th,all}} - \text{DVTSHIFT}. \tag{4.28}$$

Here, $\Delta V_{\text{th,temp}}$ is the effect of temperature on the threshold voltage and DVT-SHIFT is a parameter to model variability.

Figure 4.2 shows the validation of short-channel effects for variation in different physical parameters. Figure 4.3 shows the roll-off in V_{th} extracted from the model for different T_{ox} and compares it against V_{th} obtained from two-dimensional TCAD simulations. The smaller the T_{ox}, the closer the gate electrode is to the inversion layer, enforcing a stronger electrostatic gate control and hence smaller V_{th} roll-off.

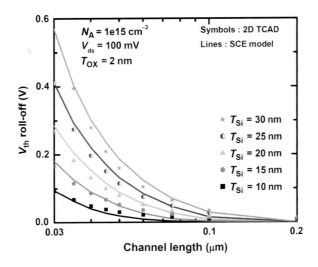

FIGURE 4.2

Threshold voltage roll-off of double-gate FETs for different T_{Si}. The model demonstrates excellent scalability when compared with two-dimensional TCAD simulations. Reproduced from [18].

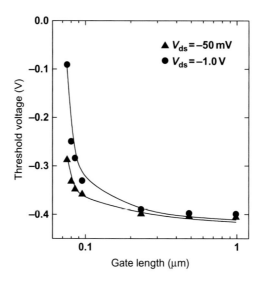

FIGURE 4.3

Model validation with experimental data of threshold voltage variation due to DIBL at $V_{ss} = -1.0\,V$ as compared with $V_{ds} = -1.0\,V$ of a lightly doped body ($N_c = 2 \times 10^{15}\,cm^{-3}$) for a p-type silicon-on-insulator (SOI) FinFET with $N_{fin} = 20$, EOT $= 2.0\,nm$, $H_{fin} = 60\,nm$, and $T_{fin} = 22\,nm$. From [20].

4.3 SUBTHRESHOLD SLOPE DEGRADATION

Short-channel effects also degrade the SS. This degradation in the SS is due to the gate bias dependence of the V_{SL} term. This is modeled similarly to the modeling in BSIM4 as follows:

$$\psi_{st} = 0.4 + \text{PHIN} + \Phi_B, \tag{4.29}$$

$$C_{dsc} = \frac{0.5}{\cosh\left(\text{DVT1SS} \cdot \frac{L_{eff}}{\lambda}\right) - 1} \cdot (\text{CDSC} + \text{CDSCD} \cdot V_{dsx}), \tag{4.30}$$

$$n = \begin{cases} \cdot\left(1 + \dfrac{\text{CIT} + C_{dsc}}{(2C_{si}) \parallel C_{ox}}\right) & \text{if GEOMOD} \neq 3, \\[3mm] \cdot\left(1 + \dfrac{\text{CIT} + C_{dsc}}{C_{ox}}\right) & \text{if GEOMOD} = 3 \text{ (nanowire)}. \end{cases} \tag{4.31}$$

This n is multiplied by the thermal voltage in the surface potential equation. Although it is an empirical formulation, it gives the desired effect as observed experimentally. Figure 4.4 shows the validation of the SS model with experimental data for different lengths in a p-type MOS FinFET.

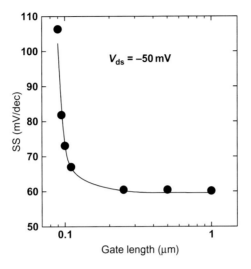

FIGURE 4.4

Model comparison with experimental data of SS for a p-type SOI FinFET with $N_{fin} = 20$, EOT = 2.0 nm, $H_{fin} = 60$ nm, and $T_{fin} = 22$ nm.

4.4 QUANTUM MECHANICAL V_{th} CORRECTION

The channel thickness in FinFETs plays an important role in the device characteristics. To control short-channel effects, the thickness of the channel is reduced. The reduction in film thickness induces a significant amount of geometrical confinement and converts these devices into a two-dimensional system. This geometrical confinement becomes more and more significant with reduction in the film thickness. It is well known from quantum mechanics that confinement brings about quantization of the energy levels. The ground-state sub-band energy level always lies above the conduction band edge. The energy difference between the conduction band edge and ground-state sub-band energy levels increases as we decrease the film thickness. This difference pushes the Fermi level up and brings it closer to the ground-state energy. This indicates that for the same bias condition, the quantum mechanical inversion charge density Q_i^{qm} will be less than the classical inversion charge density Q_i^{cl}. As a result, the quantum mechanical threshold voltage of the device becomes higher than the conventional threshold value. The change in the threshold voltage due to energy quantization in BSIM-CMG is introduced through $\Delta V_{t,QM}$. In the derivation of $\Delta V_{t,QM}$, the Schrödinger equation is solved assuming a linear potential profile for the solution of Poisson's equation across the body/channel assuming Maxwell-Boltzmann statistics in the weak region. To get accurate V_{th} for thin film thicknesses, two energy levels E_0 and E_1 are used in the $\Delta V_{t,QM}$ derivation given below. QMFACTOR also serves as a switch here.

If GEOMOD $\neq$ 3, then [17, 19]

$$E_0 = \frac{\hbar^2 \pi^2}{2m_x \cdot \text{TFIN}^2},$$ (4.32)

$$E_0' = \frac{\hbar^2 \pi^2}{2m_x' \cdot \text{TFIN}^2},$$ (4.33)

$$E_1 = 4E_0,$$ (4.34)

$$E_1' = 4E_0',$$ (4.35)

$$\gamma = 1 + \exp\left(\frac{E_0 - E_1}{kT}\right) + \frac{g' m_d'}{g m_d} \cdot \left[\exp\left(\frac{E_0 - E_0'}{kT}\right) + \exp\left(\frac{E_0 - E_1'}{kT}\right)\right],$$ (4.36)

where $\hbar$ is the reduced Planck's constant, N_c is the 3D effective density of states for the conduction band, $E_{0/1}$ ($E_{0/1}'$) is the subband energy, i.e., the separation between the 0th or 1st subband in the unprimed (primed) valley and the bottom of the conduction band at surface, and $g(g')$ and m_d (m_d') are the degeneracy and density of states effective mass in the unprimed (primed) valley, respectively.

$$\Delta V_{t,\text{QM}} = \text{QMFACTOR}_i \cdot \left[\frac{E_0}{q} - \frac{kT}{q} \ln\left(\frac{g \cdot m_d}{\pi \hbar^2 N_c} \cdot \frac{kT}{\text{TFIN}} \cdot \gamma\right)\right].$$ (4.37)

If GEOMOD $= 3$ (nanowire), then

$$E_{0,\text{QM}} = \frac{\hbar^2 (2.4048)^2}{2m_x \cdot R^2},$$ (4.38)

$$\Delta V_{t,\text{QM}} = \text{QMFACTOR} \cdot \frac{E_{0,\text{QM}}}{q}.$$ (4.39)

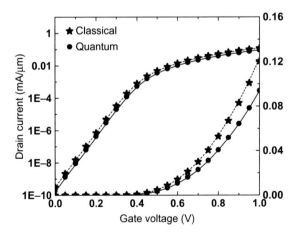

FIGURE 4.5

I_{ds}-V_{gs} comparison with and without the quantum mechanical effect.

Figure 4.5 shows a comparison of the drain current with and without the quantum mechanical effect. See also the discussion on quantum confinement in Section 4.8.

4.5 VERTICAL-FIELD MOBILITY DEGRADATION

The vertical-field-dependent mobility model of BSIM-CMG includes the mobility degradation due to different scattering mechanisms, such as phonon scattering, surface roughness scattering, and coulomb scattering. The model has been taken from BSIM4 but uses charges instead of the threshold voltage. The effect of mobility degradation is captured through the D_{mob} term in the drain current. The unified mobility model used in BSIM-CMG can be expressed as

$$\mu_{eff} = \frac{U0}{D_{mob}}, \tag{4.40}$$

where

$$D_{mob} = \begin{cases} 1 + UA \cdot (E_{eff})^{EU} + \dfrac{UD}{\left[\frac{1}{2} \cdot \left(1 + \frac{q_{ia}}{q_b}\right)\right]^{UCS}} & \text{BULKMOD} = 0, \\[2em] 1 + (UA + UC \cdot V_{eseff}) \cdot (E_{eff})^{EU} + \dfrac{UD}{\left[\frac{1}{2} \cdot \left(1 + \frac{q_{ia}}{q_b}\right)\right]^{UCS}} & \text{BULKMOD} = 1, \end{cases} \tag{4.41}$$

where

$$q_{ia} = \frac{q_{is} + q_{id}}{2}, \tag{4.42}$$

$$E_{eff} = 10^{-8} \cdot \left(\frac{q_b + \eta \cdot q_{ia}}{\epsilon_{ratio} \cdot EOT}\right), \tag{4.43}$$

$$D_{mob} = \frac{D_{mob}}{U0MULT}. \tag{4.44}$$

In Equation (4.41), UA, UC, and UD are used for phonon-surface roughness scattering, the body effect coefficient for mobility, and columbic scattering, respectively. U0MULT is a multiplier parameter used for variability modeling. and its default value is 1.0. See Appendix for description of parameters.

4.6 DRAIN SATURATION VOLTAGE, V_{dsat}

Current saturation can happen owing to pinch-off or velocity saturation or even from source-side velocity limiting. Drain saturation voltage (V_{dsat}) is needed to get the effective drain voltage V_{dseff}, which in turn is used to calculate the surface potential at the drain side (ψ_d). Note that V_{dsat} is a function of applied V_{gs} only. In other words, V_{dsat} depends only on the surface potential at the source side (ψ_s) and not on V_{ds}. Also, the presence of source/drain resistance in the transistor affects V_{dsat}. The V_{dsat} expression in BSIM-CMG is adopted from BSIM4 with the required modification. For the convenience of the reader, we present the detailed derivation of V_{dsat} for RDSMOD = 0 and RDSMOD = 1.

4.6.1 EXTRINSIC CASE (RDSMOD=1 AND 2)

In the extrinsic case, source/drain resistances are modeled using external resistors—that is, RDSMOD $= 1$ and 2 (see Chapter 7).

   When sufficiently high drain voltage is applied, the lateral electric field is large enough to cause carrier velocity to saturate near the drain end. In this case, the channel has different behavior of carrier velocities adjacent to the source and the drain ends. Adjacent to the source end, the velocity depends on the lateral field, while adjacent to the drain end, it saturates and becomes field independent. At the boundary between the two portions, the channel voltage is equal to the drain saturation voltage V_{dsat} and the lateral electric field is equal to E_{sat} [9, 10]. To use the drain current expression of BSIM4 in the derivation of V_{dsat} for BSIM-CMG, we replace $A_{\mathrm{bulk}} = 1$ and $V_{\mathrm{gsteff}} + 2\frac{kT}{q} = \mathrm{KSATIV} \cdot \left(V_{\mathrm{gsfbeff}} - \psi_s + 2\frac{kT}{q}\right)$, where KSATIV is a model parameter and has a default value of 1.0. The modified drain expression in the strong-inversion region of operation at any point y in the channel along the channel length is expressed as follows:

$$I_{\mathrm{ds}} = W_{\mathrm{eff}} \cdot C_{\mathrm{ox}} \left[\mathrm{KSATIV} \cdot \left(V_{\mathrm{gsfbeff}} - \psi_s + 2\frac{kT}{q}\right) - V(y)\right] \cdot v(y), \tag{4.45}$$

where $v(y)$ is carrier velocity at any point y in the channel and it depends upon the lateral electric field in the channel as follows,

$$v(y) = \begin{cases} \mu \cdot E_y & \text{if } E_y < E_{\mathrm{sat}}, \\ v_{\mathrm{sat}} & \text{if } E_y > E_{\mathrm{sat}}, \end{cases} \tag{4.46}$$

where μ is the mobility of the charge carriers in the channel and includes the effect of the lateral electric field as follows,

$$\mu = \frac{\mu_{\mathrm{eff}}}{1 + \frac{E(y)}{E_{\mathrm{sat}}}}. \tag{4.47}$$

The drain current before the saturation condition (i.e., in the linear or triode region of operation) in the channel can be obtained with the help of Equations (4.45) and (4.47):

$$E_y = \frac{dV(y)}{dy} = f\left(I_{\mathrm{ds}}, \psi_s, V_{\mathrm{gsfbeff}}, V_{\mathrm{ds}}\right), \tag{4.48}$$

$$\int_0^{V_{\mathrm{dsat}}} 1 \cdot dV(y) = \int_0^{L_{\mathrm{eff}}} f\left(I_{\mathrm{ds}}, \psi_s, V_{\mathrm{gsfbeff}}, V_{\mathrm{ds}}\right) \cdot dy, \tag{4.49}$$

$$I_{\mathrm{ds}} = \mu_{\mathrm{eff}} \cdot C_{\mathrm{ox}} \cdot \frac{W_{\mathrm{eff}}}{L_{\mathrm{eff}}} \cdot \frac{1}{1 + \frac{V_{\mathrm{ds}}}{E_{\mathrm{sat}} \cdot L_{\mathrm{eff}}}} \cdot \left[\mathrm{KSATIV} \cdot \left(V_{\mathrm{gsfbeff}} - \psi_s + 2\frac{kT}{q}\right) - \frac{V_{\mathrm{ds}}}{2}\right] \cdot V_{\mathrm{ds}}. \tag{4.50}$$

Now, using Equations (4.45) and (4.50) and exploiting the drain current continuity at the boundary of the linear and saturation regions of operation (at $V_{ds} = V_{dsat}$), we can write

$$\mu_{eff} \cdot C_{ox} \cdot \frac{W_{eff}}{L_{eff}} \cdot \frac{1}{1 + \frac{V_{dsat}}{E_{sat} \cdot L_{eff}}} \cdot \left[KSATIV \cdot \left(V_{gsfbeff} - \psi_s + 2\frac{kT}{q} \right) - \frac{V_{dsat}}{2} \right] \cdot V_{dsat}$$

$$= W_{eff} \cdot C_{ox} \left[KSATIV \cdot \left(V_{gsfbeff} - \psi_s + 2\frac{kT}{q} \right) - V_{dsat} \right] \cdot v_{sat}, \qquad (4.51)$$

or

$$\frac{1}{L_{eff}} \cdot \frac{1}{1 + \frac{V_{dsat}}{E_{sat} \cdot L_{eff}}} \cdot \left[KSATIV \cdot \left(V_{gsfbeff} - \psi_s + 2\frac{kT}{q} \right) - \frac{V_{dsat}}{2} \right] \cdot V_{dsat}$$

$$= \left[KSATIV \cdot \left(V_{gsfbeff} - \psi_s + 2\frac{kT}{q} \right) - V_{dsat} \right] \cdot \frac{v_{sat}}{\mu_{eff}}. \qquad (4.52)$$

In Equation (4.52), after replacing $\frac{v_{sat}}{\mu_{eff}}$ with $\frac{E_{sat}}{2}$ and $E_{sat} \cdot L_{eff}$ with E_{satL} and rearranging, we get the solution for V_{dsat} in a compact form as follows [6]:

$$V_{dsat} = \frac{E_{satL} \cdot KSATIV \cdot \left(V_{gsfbeff} - \psi_s + \frac{2kT}{q} \right)}{E_{satL} + KSATIV \cdot \left(V_{gsfbeff} - \psi_s + \frac{2kT}{q} \right)}. \qquad (4.53)$$

4.6.2 INTRINSIC CASE (RDSMOD = 0)

In the intrinsic case, part of source/drain resistances is modeled using an internal resistor—that is, RDSMOD = 0 (see Chapter 7).

As the resistance effect is included in the current equation in this case (see the D_r term in Section 4.12), V_{dsat} also changes from the intrinsic case. The series resistances cause a fraction of the applied drain voltage to drop across them. As a result, V_{dsat} increases as compared with Equation (4.53). V_{dsat} in the case of RDSMOD = 0 in BSIM-CMG is implemented in a similar way as in BSIM4 [9, 10] to reduce the complexity. The modified intrinsic drain current (without considering source/drain resistances) expression for I_{ds0} after putting $\lambda = 1$, $A_{bulk} = 1$, and $V_{gsteff} + 2\frac{kT}{q} =$ KSATIV $\cdot \left(V_{gsfbeff} - \psi_s + 2\frac{kT}{q} \right)$ in the BSIM4 model can be written as follows:

$$I_{ds0} = \frac{W_{eff} \cdot C_{ox} \cdot KSATIV \cdot \left(V_{gsfbeff} - \psi_s + 2\frac{kT}{q} \right)}{\left(1 + \frac{V_{ds}}{E_{sat} \cdot L_{eff}} \right)} \cdot \mu_{eff} \cdot V_{ds}$$

$$\times \left(1 - \frac{V_{ds}}{2 \cdot KSATIV \cdot \left(V_{gsfbeff} - \psi_s + 2\frac{kT}{q} \right)} \right). \qquad (4.54)$$

Here, KSATIV is a model parameter, and it has a default value of 1.0. After considering the source/drain resistance $R_{ds}(V)$ and applying Ohm's law, we can write the drain current as follows:

$$I_{ds} = \frac{V_{ds}}{R_{ch} + R_{ds}},$$ (4.55)

or

$$I_{ds} = \frac{I_{ds0}}{1 + \frac{R_{ds} \cdot I_{ds0}}{V_{ds}}},$$ (4.56)

where $R_{ch} = \frac{V_{ds}}{I_{ds0}}$ when $R_{ds} = 0$. The drain current in the saturation region of operation can be expressed as follows:

$$I_{ds,sat} = W_{eff} \cdot q_{ch} \, (V(y) = V_{dsat}) \cdot \text{VSAT},$$ (4.57)

$$I_{ds,sat} = W_{eff} \cdot C_{ox} \cdot \text{KSATIV} \cdot \left(V_{gsfbeff} - \psi_s + 2\frac{kT}{q} \right) \cdot \text{VSAT}$$

$$\times \left(1 - \frac{V_{dsat}}{\text{KSATIV} \cdot \left(V_{gsfbeff} - \psi_s + 2\frac{kT}{q} \right)} \right).$$ (4.58)

At $V_{ds} = V_{dsat}$, the channel current expressed by Equation (4.56) will be equal to the drain current in Equation (4.58) (i.e., $I_{ds,sat}$):

$$\frac{I_{ds0}(V_{dsat})}{1 + \frac{R_{ds} \cdot I_{ds0}}{V_{dsat}}} = W_{eff} \cdot C_{ox} \cdot \text{KSATIV} \cdot \left(V_{gsfbeff} - \psi_s + 2\frac{kT}{q} \right) \cdot \text{VSAT}$$

$$\times \left(1 - \frac{V_{dsat}}{\text{KSATIV} \cdot \left(V_{gsfbeff} - \psi_s + 2\frac{kT}{q} \right)} \right).$$ (4.59)

After putting the value of I_{ds0} from Equation (4.56) in Equation (4.59), we can write the equation above as follows:

$$\frac{\frac{W_{eff} \cdot C_{ox} \cdot \text{KSATIV} \cdot \left(V_{gsfbeff} - \psi_s + 2\frac{kT}{q} \right)}{\left(1 + \frac{V_{dsat}}{E_{sat} \cdot L_{eff}} \right)} \cdot \mu_{eff} \cdot V_{dsat} \times \left(1 - \frac{V_{dsat}}{2 \cdot \text{KSATIV} \cdot \left(V_{gsfbeff} - \psi_s + 2\frac{kT}{q} \right)} \right)}{1 + \frac{R_{ds}}{V_{dsat}} \cdot \frac{W_{eff} \cdot C_{ox} \cdot \text{KSATIV} \cdot \left(V_{gsfbeff} - \psi_s + 2\frac{kT}{q} \right)}{\left(1 + \frac{V_{dsat}}{E_{sat} \cdot L_{eff}} \right)} \cdot \mu_{eff} \cdot V_{dsat} \times \left(1 - \frac{V_{dsat}}{2 \cdot \text{KSATIV} \cdot \left(V_{gsfbeff} - \psi_s + 2\frac{kT}{q} \right)} \right)}$$

$$= W_{eff} \cdot C_{ox} \cdot \text{KSATIV} \cdot \left(V_{gsfbeff} - \psi_s + 2\frac{kT}{q} \right) \cdot \text{VSAT}$$

$$\times \left(1 - \frac{V_{dsat}}{\text{KSATIV} \cdot \left(V_{gsfbeff} - \psi_s + 2\frac{kT}{q} \right)} \right).$$ (4.60)

In Equation (4.60), if we put $\mu_{eff} = \frac{2 \cdot \text{VSAT}}{E_{sat}}$ and $E_{sat} \cdot L_{eff} = E_{satL}$ and after some manipulation and rearranging, we get the following form of Equation (4.60):

$$\frac{V_{dsat} \cdot \left[2 \cdot \text{KSATIV} \cdot \left(V_{gsfbeff} - \psi_s + 2\frac{kT}{q}\right) - V_{dsat}\right]}{E_{satL} + V_{dsat} + W_{eff} \cdot C_{ox} \cdot \text{VSAT} \cdot \left[2 \cdot \text{KSATIV} \cdot \left(V_{gsfbeff} - \psi_s + 2\frac{kT}{q}\right) - V_{dsat}\right] \cdot R_{ds}}$$

$$= \left[\text{KSATIV} \cdot \left(V_{gsfbeff} - \psi_s + 2\frac{kT}{q}\right) - V_{dsat}\right], \tag{4.61}$$

$$W_{eff} \cdot C_{ox} \cdot \text{VSAT} \cdot R_{ds} \cdot V_{dsat}^2 - 3W_{eff} \cdot C_{ox} \cdot \text{VSAT} \cdot R_{ds} \cdot \text{KSATIV} \cdot \left(V_{gsfbeff} - \psi_s + 2\frac{kT}{q}\right)$$

$$\times V_{dsat} - \left[\text{KSATIV} \cdot \left(V_{gsfbeff} - \psi_s + 2\frac{kT}{q}\right) + E_{satL}\right] \cdot V_{dsat}$$

$$\times 2W_{eff} \cdot C_{ox} \cdot \text{VSAT} \cdot \text{KSATIV} \cdot \left(V_{gsfbeff} - \psi_s + 2\frac{kT}{q}\right)^2$$

$$+ E_{satL} \cdot \text{KSATIV} \cdot \left(V_{gsfbeff} - \psi_s + 2\frac{kT}{q}\right) = 0. \tag{4.62}$$

Equation (4.62) is now converted into the standard form of $a \cdot x^2 - b \cdot x + c = 0$, and it can be written as follows:

$$\frac{a}{2} \cdot V_{dsat}^2 - b \cdot V_{dsat} + c = 0, \tag{4.63}$$

where

$$a = 2 \cdot W_{eff} \text{VSAT} C_{ox} \cdot R_{ds}, \tag{4.64}$$

$$b = E_{satL} + \text{KSATIV} \left(V_{gsfbeff} - \psi_s + 2\frac{kT}{q}\right)\left(1 + \frac{3}{2}T_a\right), \tag{4.65}$$

$$c = \text{KSATIV} \left(V_{gsfbeff} - \psi_s + 2\frac{kT}{q}\right)\left[E_{satL} + T_a \cdot \text{KSATIV} \cdot \left(V_{gsfbeff} - \psi_s + 2\frac{kT}{q}\right)\right]. \tag{4.66}$$

V_{dsat} in terms of T_a, T_b, and T_c can then be written as follows:

$$V_{dsat} = \frac{T_b - \sqrt{(T_b)^2 - 2 \cdot T_a T_c}}{T_a}. \tag{4.67}$$

Once V_{dsat} has been evaluated, V_{dseff} is calculated using the following formula:

$$V_{dseff} = \frac{V_{ds}}{\left[1 + \left(\frac{V_{ds}}{V_{dsat}}\right)^{\text{MEXP}}\right]^{\frac{1}{\text{MEXP}}}}. \tag{4.68}$$

The interpolation function used in V_{dseff} for a smooth transition from V_{ds} to V_{dsat} ensures symmetry around $V_{dseff} = 0$. Another beauty of this function is that it forces

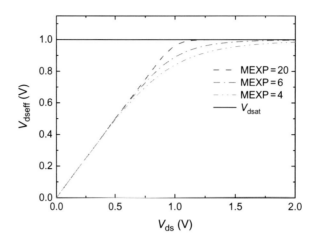

FIGURE 4.6

The effective drain-source voltage V_{dseff} (solid lines) versus V_{ds} for various values of MEXP.

V_{dseff} to be zero when $V_{ds} = 0$ without being affected, by the compiler's numerical precision (Figure 4.6).

4.7 VELOCITY SATURATION MODEL

The effect of velocity saturation on the drain current in BSIM-CMG is included through D_{vsat}. The velocity saturation model in BSIM-CMG is similar to that for BSIM6 bulk MOSFETs [11, 12]:

$$E_{sat1} = \frac{2 \cdot \text{VSAT1} \cdot D_{mob}}{\mu_0}, \qquad (4.69)$$

where VSAT1 (= VSAT by default) has been provided for extra flexibility for separate tuning of V_{dsat} and current.

$$D_{vsat} = \frac{1 + \left[\text{DELTAVSAT} + \left(\frac{\Delta q_i}{E_{sat1} \cdot L_{eff}} \right)^{\text{PSAT}} \right]^{\frac{1}{\text{PSAT}}}}{1 + \text{DELTAVSAT}^{\frac{1}{\text{PSAT}}}} + \frac{1}{2} \cdot \text{PTWG} \cdot q_{ia} \cdot \Delta \psi^2, \qquad (4.70)$$

where DELTAVSAT, PSAT, and PTWG are velocity saturation parameters. D_{mob} is the vertical-field-induced mobility degradation given by Equation (4.41). The default value of the PSAT parameter is 2. In Equation (4.70), the second term containing $\Delta \psi^2$ is added empirically to have more flexibility in the model for better fitting of the g_{msat} behavior.

Some devices do not exhibit prominent or abrupt velocity saturation. The parameters A1 and A2 are used to tune this nonsaturation effect to fit I_{dsat} or g_{msat}:

$$T_0 = \max \left[\left(A1 + \frac{A2}{q_{ia} + 2.0 \cdot \frac{nkT}{q}} \right) \cdot \Delta q_i^2, -1 \right], \tag{4.71}$$

$$N_{sat} = \frac{1 + \sqrt{1 + T0}}{2}, \tag{4.72}$$

$$D_{vsat} = D_{vsat} \cdot N_{sat}. \tag{4.73}$$

4.8 QUANTUM MECHANICAL EFFECTS

The quantum mechanical effects in multigate field effect devices are more influential in the device characteristics than in conventional bulk MOSFETs. Multigate devices have additional effects arising from geometrical confinement along with the electrical confinement. It is well known that quantum mechanical effects cause the average charge location to shift away from the interface (called charge centroid shift) [14]. In BSIM-CMG, the threshold voltage shift due to bias-dependent ground-state sub-band energy is considered in the surface potential calculation as given by Equation (4.32). The charge centroid shift in BSIM-CMG is modeled through variation of the oxide thickness and width [13] and the quantum mechanical effect is activated by the parameters QMTCENIV and QMTCENCV for *I-V* and *C-V* characteristics, respectively. The charge centroid in the inversion region can be expressed as [10, 13, 14]

$$T_{cen} = \frac{T_{cen0}}{1 + \left(\frac{q_{ia} + ETAQM \cdot q_{ba}}{QM0} \right)^{PQM}}, \tag{4.74}$$

where ETAQM, QM0, and PQM are parameters which vary the location of the charge centroid with applied bias. T_{cen0} used in the equation above is calculated from geometrical parameters—that is, the fin height and fin width in the case of a double-gate FET and a triple-gate FET and the radius of nanowire in the case of nanowire. With sue of coupled Poisson-Schrödinger equations, it has been shown that T_{cen0} varies from $0.25T_{fin}$ for a thick fin to $0.351T_{fin}$ for a thin fin in a double-gate FET. To model this dependence, an empirical function is used in BSIM-CMG, and is given by [13]

$$\frac{T_{cen0}}{T_{fin}} = 0.25 + (0.351 - 0.25) \exp \left(\frac{T_{fin}}{T_0} \right), \tag{4.75}$$

where T_0 is a fitting parameter. Similarly, nanowire has the following relation between T_{cen0} and the nanowire radius R:

$$\frac{T_{cen0}}{R} = 0.334 + (0.576 - 0.334) \exp\left(\frac{R}{R_0}\right), \tag{4.76}$$

where R_0 is a fitting parameter. The functions used for T_{cen0} are validated for different thicknesses and radii with TCAD data and show an excellent match as shown in Figure 4.7. The variation of charge centroid with gate voltage (or inversion charge) given by Equation (4.74) is also validated with TCAD simulations as shown in Figure 4.8. After a certain critical charge in the channel, the centroid shifts toward the gate-oxide channel interface with increasing gate bias (Figure 4.9).

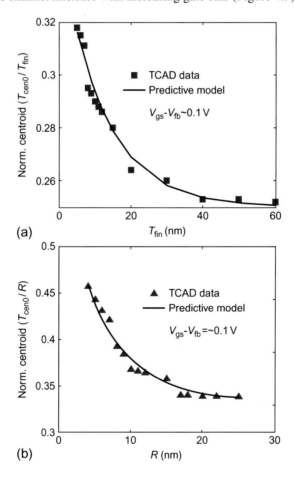

FIGURE 4.7

Dependency of the charge centroid on channel thickness for a double-gate FET and a nanowire FET at low gate voltages [13].

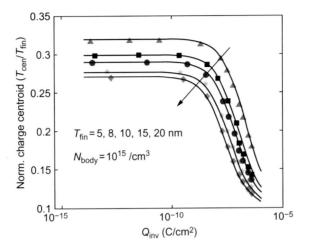

FIGURE 4.8

Shifting of the charge centroid with inversion charge in the channel. The figure shows shifting of the charge centroid toward the Si-SiO$_2$ interface after a critical charge value depending upon the channel thickness [13].

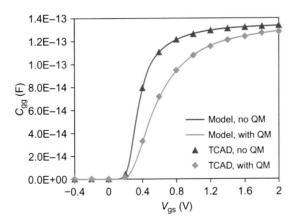

FIGURE 4.9

Small-signal gate capacitance (C_{gg}) versus gate voltage for a long-channel double-gate FET with $T_{fin} = 15$ nm and $T_{ox} = 1.5$ nm [13].

The shifting of the centroid causes the width to vary with the applied bias—that is, the width increases as centroid shifts toward interface with an increase in gate voltage, saturating in deep strong inversion. The effective width for *I-V* characteristics is calculated from QMTCENIV and T_{cen}. Similarly, the effective width and effective oxide thickness for *C-V* characteristics are calculated from QMTCENCV and T_{cen}.

4.8.1 EFFECTIVE WIDTH MODEL

The expressions for the effective width for different multigate structures are given below, and these are used in current and capacitance modules:

$$W_{\text{eff}} = W_{\text{eff0}} - \Lambda \cdot \text{QMTCENIV} \cdot T_{\text{cen}}, \tag{4.77}$$

$$W_{\text{eff,CV}} = W_{\text{eff,CV0}} - \Lambda \cdot \text{QMTCENCV} \cdot T_{\text{cen}}, \tag{4.78}$$

where

$$\Lambda = 0 \text{ for a double-gate FET (GEOMOD} = 0), \tag{4.79}$$

$$\Lambda = 4 \text{ for a triple-gate FET (GEOMOD} = 1), \tag{4.80}$$

$$\Lambda = 8 \text{ for a quadruple-gate FET (GEOMOD} = 2), \tag{4.81}$$

$$\Lambda = 2\pi \text{ for a surround-gate FET (GEOMOD} = 3). \tag{4.82}$$

In these calculations, the effective width changes from W_{eff0} and $W_{\text{eff,CV0}}$ only when QMTCENIV and QMTCENCV, respectively, are *nonzero*. By default, *C-V* calculations use the same parameter values as used in the *I-V* model, unless the value of QMTCENCV is provided separately in the model card.

4.8.2 EFFECTIVE OXIDE THICKNESS/EFFECTIVE CAPACITANCE

A quantum mechanical effect on *C-V* characteristics is enabled by a nonzero value of QMTCENCV. When QMTCENCV $\neq 0$, the effective insulator capacitance $C_{\text{ox,eff}}$ is used for the *C-V* calculation but *I-V* characteristics always use C_{ox}. For $C_{\text{ox,eff}}$ evaluation, T_{cen} is added to the scaled physical oxide thickness (TOXP) to account for any dielectric material.

If QMTCENCV $\neq 0$, then

$$C_{\text{ox,eff}} = \begin{cases} \dfrac{3.9 \cdot \epsilon_0}{\text{TOXP}\frac{3.9}{\text{EPSROX}} + T_{\text{cen}} \cdot \frac{\text{QMTCENCV}}{\epsilon_{\text{ratio}}}} & \text{GEOMOD} \neq 3, \\[4mm] \dfrac{3.9 \cdot \epsilon_0}{R \cdot \left[\frac{1}{\epsilon_{\text{ratio}}} \ln\left(\frac{R}{R - T_{\text{cen}}}\right) + \frac{3.9}{\text{EPSROX}} \ln\left(1 + \frac{\text{TOXP}}{R}\right)\right]} & \text{GEOMOD} = 3. \end{cases} \tag{4.83}$$

If QMTCENCV $= 0$, then

$$C_{\text{ox,eff}} = C_{\text{ox}}. \tag{4.84}$$

4.8.3 CHARGE CENTROID CALCULATION FOR ACCUMULATION

If QMTCENCV $\neq 0$, then

$$
C_{\text{ox,acc}} = \begin{cases}
\dfrac{3.9 \cdot \epsilon_0}{\text{TOXP}\frac{3.9}{\text{EPSROX}} + \dfrac{T_{\text{cen0}}}{1 + \left(\frac{q_{\text{i,acc}}}{\text{QM0ACC}}\right)^{\text{PQMACC}}} \cdot \dfrac{\text{QMTCENCVA}}{\epsilon_{\text{ratio}}}} & \text{GEOMOD} \neq 3, \\[3em]
\dfrac{3.9 \cdot \epsilon_0}{\left[R \cdot \left[\dfrac{1}{\epsilon_{\text{ratio}}} \ln\left(\dfrac{R}{R - \dfrac{T_{\text{cen0}}}{1 + \left(\frac{q_{\text{i,acc}}}{\text{QM0ACC}}\right)^{\text{PQMACC}}}} \right) + \dfrac{3.9}{\text{EPSROX}} \ln\left(1 + \frac{\text{TOXP}}{R}\right) \right] \right]} & \text{GEOMOD} = 3,
\end{cases}
$$

$$(4.85)$$

If QMTCENCV $= 0$, then

$$
C_{\text{ox,acc}} = \frac{3.9 \cdot \epsilon_0}{\text{EOTACC}}. \tag{4.86}
$$

4.9 LATERAL NONUNIFORM DOPING MODEL

Lateral nonuniform doping along the length of the channel leads to *I-V* and *C-V* displaying different threshold voltages. However, the consistent-surface-potential-based *I-V* and *C-V* model does not allow for the use of different V_{th} values. A straightforward method would be to recompute the surface potentials at the source and drain ends twice for *I-V* and *C-V* separately, breaking the consistency but at the expense of computation time. The following model has been introduced as a multiplicative factor to the drain current (*I-V*) to allow for that V_{th} shift:

$$
M_{\text{nud}} = e^{\left(-\frac{\text{K0}(T)}{\text{K0SI}(T) \cdot q_{\text{ia}} + 2.0 \cdot \frac{nkT}{q}}\right)}. \tag{4.87}
$$

This model should be applied after the *C-V* extraction step to match the V_{th} for the subthreshold region in the $I_{\text{d(lin)}}$-V_{g} curve. The parameter K0 is used to fit the subthreshold region, while the parameter K0SI helps reclaim the fit in the inversion region.

4.10 BODY EFFECT MODEL FOR A BULK FinFET (BULKMOD = 1)

A small amount of body bias effect can be captured from the following model. This model is applicable only for *I-V* and not for *C-V*.

$$
V_{\text{esx}} = V_{\text{es}} - 0.5 \cdot (V_{\text{ds}} - V_{\text{dsx}}), \tag{4.88}
$$

$$
V_{\text{eseff}} = \min(V_{\text{esx}}, 0.95 \cdot \text{PHIBE}_{\text{i}}), \tag{4.89}
$$

$$
dV_{\text{thBE}} = \sqrt{\text{PHIBE}_{\text{i}} - V_{\text{eseff}}} - \sqrt{\text{PHIBE}_{\text{i}}}, \tag{4.90}
$$

$$M_{ob} = \exp\left(-dV_{thBE}\frac{K1(T) + K1SAT(T) \cdot V_{dsx}}{K1SI(T) \cdot q_{ia} + 2.0 \cdot \frac{nkT}{q}}\right). \tag{4.91}$$

The lateral nonuniform doping model and the body effect model are empirical and have their limits as to how much V_{th} shift can be achieved without distorting the I-V curve. Overusage could lead to negative g_m or negative g_{ds}. The lateral nonuniform doping model could be used in combination with the mobility model to achieve high V_{th} shift between C-V and I-V curves to avoid any distortion of higher-order derivatives.

4.11 OUTPUT RESISTANCE MODEL

Analog behavior of a transistor, to a large extent, depends on the output resistance as gain is proportional to $\frac{g_m}{g_{ds}}$. Therefore, it is extremely important to correctly model it for different biases and dimensions. The continuous shrinking of geometrical dimensions has resulted in a significant increase in short-channel effects on the output resistance and has posed additional modeling challenges to model engineers. Earlier compact models accounted only for CLM and therefore were inaccurate. The first comprehensive analytical scalable compact model of output resistance of a bulk MOSFET was proposed by Huang et al. [15].

Figure 4.10 shows the variation of typical on resistance of a short-channel bulk MOSFET with drain bias. As indicated in the figure, different short-channel effects,

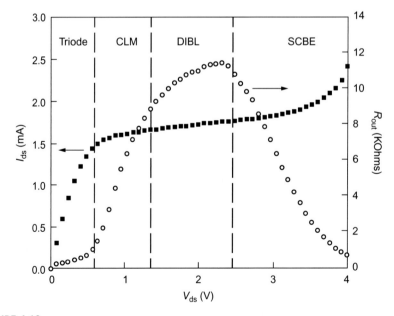

FIGURE 4.10

Drain current and on resistance variation with drain bias [15].

affect different regions of the characteristics. BSIM-CMG employs an early voltage (V_A) based model for the output conductance, similarly to BSIM4 and BSIM6 [10, 12, 15].

Drain current is a weak function of drain voltage in the saturation region. It can be approximated as

$$I_{ds}(V_{gs}, V_{ds}) = I_{ds}(V_{gs}, V_{dsat}) + \frac{dI_{ds}(V_{gs}, V_{ds})}{dV_{ds}}(V_{ds} - V_{dsat}) \tag{4.92}$$

$$= I_{dsat}\left(1 + \frac{V_{ds} - V_{dsat}}{V_A}\right), \tag{4.93}$$

where

$$I_{dsat} = I_{ds}(V_{gs}, V_{dsat}), \tag{4.94}$$

$$V_A = \frac{I_{dsat}}{\left(\frac{dI_{ds}}{dV_{ds}}\right)}. \tag{4.95}$$

4.11.1 CHANNEL-LENGTH MODULATION

Figure 4.11 shows that in saturation, the MOSFET channel can be visualized as composed of two parts—one extending from the source to the saturation point and the other extending from the saturation point to the drain. The saturation point of the channel is a function of drain bias and shifts toward the source as drain bias is increased. As a result, the effective channel length reduces and the drain current increases with drain voltage even in saturation. The reduction in channel length is an increasing function of V_{ds}-V_{dsat}. This effect is called CLM. The same arguments can safely be extended to multigate transistors.

As shown in Figure 4.12, the inversion layer near the drain is located below the surface, and accurate modeling of CLM requires two-dimensional analysis in that region. The channel-length reduction is given by [9, 16, 17]

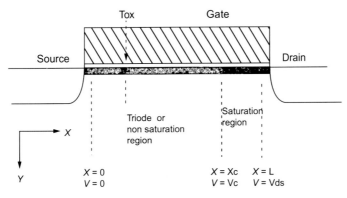

FIGURE 4.11

Representation of a bulk MOSFET in the saturation region. Reproduced from [21].

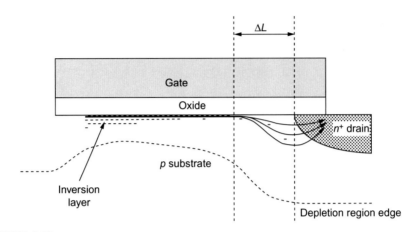

FIGURE 4.12

Inversion layer in saturation in a bulk MOSFET. Reproduced from [22].

$$\Delta L = \text{Litl} \cdot \sinh^{-1}\left(\frac{V_{ds} - V_{dsat}}{E_{sat} \cdot \text{Litl}}\right), \tag{4.96}$$

where the characteristic drain field length (Litl) is given by

$$\text{Litl} = \sqrt{\frac{\epsilon \cdot \text{TOXE} \cdot \text{XJ}}{\text{EPSROX}}}. \tag{4.97}$$

Early voltage due to CLM is given by

$$V_{A,CLM} = \frac{I_{dsat}}{\left(\frac{dI_{ds}}{dL} \cdot \frac{dL}{dV_{ds}}\right)}. \tag{4.98}$$

From quasi-two-dimensional analysis [16], it can be equivalently written as

$$V_{A,CLM} = C_{clm} \cdot (V_{ds} - V_{dsat}), \tag{4.99}$$

$$\frac{1}{C_{clm}} = \begin{cases} \text{PCLM} + \text{PCLMG} \cdot q_{ia} & \text{for PCLMG}_i > 0, \\ \dfrac{1}{\frac{1}{\text{PCLM}} - \text{PCLMG} \cdot q_{ia}} & \text{for PCLMG}_i < 0, \end{cases} \tag{4.100}$$

$$M_{clm} = 1 + \frac{1}{C_{clm}} \ln\left(1 + \frac{V_{ds} - V_{dseff}}{V_{dsat} + E_{satL}} \cdot C_{clm}\right). \tag{4.101}$$

M_{clm} is multiplied by I_{ds} in the final drain current expression.

4.11.2 DRAIN-INDUCED BARRIER LOWERING

DIBL is another important short-channel effect, and significantly affects output resistance. For a short-channel device, applied drain voltage reduces the barrier between the source and the channel, allowing more carriers to enter the channel compared with the number predicted from long-channel theory. This manifests itself as a reduction in the threshold voltage and an increase in drain current. The larger the drain bias, the more prominent is the barrier lowering, and thus current becomes a function of the drain voltage.

Although the DIBL effect is already captured through the shift of the threshold voltage (see Equation 4.25), it is not sufficient for accurate fitting of current and output resistance in saturation. Additional flexibility is added in the model using the early voltage concept for DIBL modeling [15].

$$
\text{PVAG factor} = \begin{cases} 1 + \text{PVAG} \cdot \frac{q_{ia}}{E_{sat}L_{eff}} & \text{for PVAG} > 0, \\ \frac{1}{1 - \text{PVAG} \cdot \frac{q_{ia}}{E_{sat}L_{eff}}} & \text{for PVAG} < 0, \end{cases} \tag{4.102}
$$

$$
\theta_{rout} = \frac{0.5 \cdot \text{PDIBL1}}{\cosh\left(\text{DROUT} \cdot \frac{L_{eff}}{\lambda}\right) - 1} + \text{PDIBL2}, \tag{4.103}
$$

$$
V_{ADIBL} = \frac{q_{ia} + 2kT/q}{\theta_{rout}} \cdot \left(1 - \frac{V_{dsat}}{V_{dsat} + q_{ia} + 2kT/q}\right) \cdot \text{PVAG factor}, \tag{4.104}
$$

$$
M_{oc} = \left(1 + \frac{V_{ds} - V_{dseff}}{V_{ADIBL}}\right) \cdot M_{clm}. \tag{4.105}
$$

M_{oc} is multiplied by I_{ds} in the final drain current expression.

4.12 CHANNEL CURRENT

The drain-to-source current including all effects is given by

$$
I_{ds} = \text{IDS0MULT} \cdot \mu_0 \cdot C_{ox} \cdot \frac{W_{eff}}{L_{eff}} \cdot i_{ds0} \cdot \frac{M_{oc}M_{ob}M_{nud}}{D_{mob} \cdot D_r \cdot D_{vsat}} \times \text{NFIN}_{total}, \tag{4.106}
$$

where i_{ds0} is the core current model developed in Section 3.1. IDS0MULT is a multiplier for the drain current and is used for variability modeling. The "D_r" term is the effect of source-drain resistance on channel currently given as follows for different RDSMOD cases (see Chapter 7 for more details).

RDSMOD = 0 (**Internal bias dependent, external bias independent**)

$$
R_{source} = R_{s,geo} \tag{4.107}
$$

$$
R_{drain} = R_{d,geo} \tag{4.108}
$$

$$
R_{ds} = \frac{1}{\text{NFIN}_{total} \times \text{Weff0}^{WR}} \cdot \left(\text{RDSWMIN} + \frac{\text{RDSW}}{1 + \text{PRWGS} \cdot q_{ia}}\right) \tag{4.109}
$$

$$
D_r = 1.0 + \text{NFIN}_{total} \times \mu_0 \cdot C_{ox} \cdot \frac{W_{eff}}{L_{eff}} \cdot \frac{i_{ds0}}{\Delta q_i} \cdot \frac{R_{ds}}{D_{vsat} \cdot D_{mob}}
$$

$\underline{RDSMOD} = 1$ **(External)**

$$D_r = 1.0 \tag{4.110}$$

$\underline{RDSMOD} = 2$ **(Internal bias independent and bias dependent)**

$$R_{source} = 0.0 \tag{4.111}$$

$$R_{drain} = 0.0 \tag{4.112}$$

$$R_{ds} = \frac{1}{\text{NFIN}_{total} \times \text{Weff0}^{WR}} \cdot \left(R_{s,geo} + R_{d,geo} + \text{RDSWMIN} + \frac{\text{RDSW}}{1 + \text{PRWGS} \cdot q_{ia}} \right) \tag{4.113}$$

$$D_r = 1.0 + \text{NFIN}_{total} \times \mu_0 \cdot C_{ox} \cdot \frac{W_{eff}}{L_{eff}} \cdot \frac{i_{ds0}}{\Delta q_i} \cdot \frac{R_{ds}}{D_{vsat} \cdot D_{mob}}$$

$R_{s,geo}$ and $R_{d,geo}$ are the source and drain diffusion resistances.

REFERENCES

[1] K. Suzuki, T. Tanaka, Y. Tosaka, H. Horie, Y. Arimoto, Scaling theory for double-gate SOI MOSFETs, IEEE Trans. Electron Devices 40 (12) (1993) 2326–2329.

[2] K. Suzuki, Y. Tosaka, T. Sugii, Analytical threshold voltage model for short channel n^+/p^+ double-gate SOI MOSFETs, IEEE Trans. Electron Devices 43 (5) (1996) 732–738.

[3] G. Pei, J. Kedzierski, P. Oldiges, M. Ieong, E. Kan, FinFET design considerations based on 3-D simulation and analytical modeling, IEEE Trans. Electron Devices 48 (8) (2002) 1441–1419.

[4] C.-H. Lin, Compact modeling of nanoscale CMOS (Ph.D. Dissertation), University of California, Berkeley, 2007.

[5] Z.H. Liu, C. Hu, J.H. Huang, T.Y. Chan, M.C. Jeng, P.K. Ko, Y.C. Cheng, Threshold voltage model for deep-submicrometer MOSFETs, IEEE Trans. Electron Devices 40 (1) (1993) 86–95.

[6] C.C. Hu, Modern Semiconductor Devices for Integrated Circuits, Prentice Hall, Upper Saddle River, NJ, 2010.

[7] Y. Tsividis, C. McAndrew, Operation and modeling of the MOS transistor, in: Oxford Series in Electrical and Computer Engineering, 2010.

[8] K. Cao, W. Liu, X. Jin, K. Vasanth, K. Green, J. Krick, T. Vrotsos, C. Hu, Modeling of pocket implanted MOSFETs for anomalous analog behavior, in: IEDM Tech. Dig., 1999, pp. 171–174.

[9] BSIM4 Technical Manual and Code. Available from: http://www-device.eecs.berkeley.edu/bsim/?page=BSIM4.

[10] W. Liu, C. Hu, BSIM4 and MOSFET Modeling for IC Simulation, World Scientific, Singapore, 2011.

[11] Y.S. Chauhan, S. Venugopalan, M.-A. Chalkiadaki, M.A. Karim, H. Agarwal, S. Khandelwal, N. Paydavosi, J.P. Duarte, C.C. Enz, A.M. Niknejad, C. Hu, BSIM6: analog and RF compact model for bulk MOSFET, IEEE Trans. Electron Devices 61 (2) (2014) 234–244.

[12] BSIM6 Technical Manual and Code. Available from: http://www-device.eecs.berkeley.edu/bsim/?page=BSIM6.

[13] S. Venugopalan, M.A. Karim, S. Salahuddin, A.M. Niknejad, C.C. Hu, Phenomenological compact model for QM charge centroid in multigate FETs, IEEE Trans. Electron Devices 60 (4) (2013) 1480–1484.

[14] W. Liu, X. Jin, Y. King, C. Hu, An efficient and accurate compact model for thin-oxide-MOSFET intrinsic capacitance considering the finite charge layer thickness, IEEE Trans. Electron Devices 46 (5) (1999) 1070–1072.

[15] J.H. Huang, Z.H. Liu, M.C. Jeng, P.K. Ko, C. Hu, A physical model for MOSFET output resistance, in: IEEE International Electron Device Meeting, 1992, pp. 569–572.

[16] N.G. Einspruch, G. Gildenblat, Advanced MOS Device Physics. Academic Press, New York, 1989.

[17] BSIM-CMG Technical Manual and Code. Available from: http://www-device.eecs.berkeley.edu/bsim/.

[18] M.V. Dunga, Nanoscale CMOS modeling, Ph.D. dissertation, University of California, Berkeley, 2008.

[19] V.P. Trivedi, J.G. Fossum, Quantum-mechanical effects on the threshold voltage of undoped double-gate MOSFETs, IEEE Electron Device Lett. 26 (8) (2005) 579–582.

[20] Y.S. Chauhan, D.D. Lu, S. Venugopalan, M.A. Karim, A. Niknejad, C. Hu, Compact models for sub-22 nm MOSFETs, Workshop on Compact Modeling, Boston, USA, June 2011.

[21] Y. Cheng, C. Hu, MOSFET Modeling and BSIM3 User's Guide, Kluwer Academic Publishers, 2002.

[22] Y. Tsividis, C. McAndrew, Operation and Modeling of the MOS Transistor, Oxford University Press, 2011.

Leakage currents

5

CHAPTER OUTLINE

5.1 Weak-inversion current 129
5.2 Gate-induced source and drain leakages 130
    5.2.1 GIDL/GISL current formulation in BSIM-CMG 132
5.3 Gate oxide tunneling 133
    5.3.1 Gate oxide tunneling formulation in BSIM-CMG 134
    5.3.2 Gate-to-body tunneling current in depletion/inversion 135
    5.3.3 Gate-to-body tunneling current in accumulation 136
    5.3.4 Gate-to-channel tunneling current in inversion 137
    5.3.5 Gate-to-source/drain tunneling current 138
5.4 Impact ionization 140
References 141

During the past four decades, while IC manufacturers were continually reducing the physical size of planar silicon MOSFETs in order to improve their speed and power efficiency and to lower the fabrication cost per transistor, an undesirable effect was growing in parallel. Because of short-channel effects, the leakage current and, consequently, the leakage (static) power dissipation were increasing. Today, conventional planar, bulk MOSFET scaling has reached a point where almost half of the power that is dissipated in a chip is due to the static leakage power. The conventional bulk MOSFET scaling is coming to an end not because of fabrication difficulties, rather because of the fact that further scaling would not decrease power dissipation and may in fact increase it. As described in Chapter 1, the FinFET architecture could greatly reduce the short-channel effects, convincing the industry to alter the architecture of the traditional MOSFET from planar to FinFET.

If we plot the drain current I_d of a typical MOSFET (on a logarithmic scale) as a function of its gate voltage V_{gs} for a nonzero drain voltage $V_{ds} \neq 0$, the value of the intersection with the drain current axis gives the off-state leakage current (see Figure 5.1). Typically, the transistor's off-state leakage current for V_{ds} equal to the supply voltage V_{dd} is defined as the off-state current, I_{off}. Ideally, the transistor is

FinFET Modeling for IC Simulation and Design. http://dx.doi.org/10.1016/B978-0-12-420031-9.00005-1

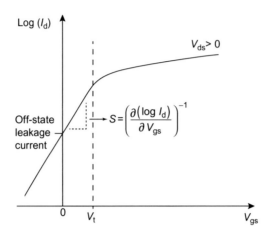

FIGURE 5.1

The transistor's drain current (on a logarithmic scale) as a function of its gate voltage. The off-state leakage current, threshold voltage, and subthreshold swing are marked.

supposed to be completely off at this bias point, but as explained later in this chapter, there always exist leakage currents due to different mechanisms. These sources of leakage current include the weak-inversion current between the drain terminal and source terminal, the substrate and drain junction leakage currents (both forward[1] and reverse diode currents), the gate-induced drain leakage (GIDL) current between the drain terminal and the substrate terminal, and a portion of the gate oxide tunneling current.

From those, the junction leakage and gate oxide tunneling currents extend to the transistor's on state and add to impact ionization leakage, which becomes noticeable in the on state. In addition, there might be leakage currents between terminal pairs other than those involving the drain; for instance, gate-induced source leakage (GISL) between the source and substrate terminals. BSIM-CMG is equipped with models that can simulate the leakage currents of all the terminals.[2]

In Section 5.1, the weak-inversion current is reviewed, with a focus on the terminology used in the field. In Section 5.2, an approach similar to the one originally used in BSIM4 is described to develop a GIDL/GISL model for BSIM-CMG. Section 5.3 discusses the gate oxide tunneling mechanisms and formulations. Finally, Section 5.4 explains the impact ionization model. The junction leakage component is reviewed in detail in Chapter 9.

---

[1]Forward leakage can occur under intentional forward well bias and under voltage spikes that may even lead to latch-up.

[2]Note that for a FinFET on a silicon-on-insulator (SOI) substrate (BULKMOD = 0), the substrate leakage current will flow out of the source; that is, the holes will be injected into the source and appear as an additional drain-source leakage.

5.1 **WEAK-INVERSION CURRENT**

Assuming a room-temperature n-channel MOSFET with $V_{gs} < V_t$, there are always some electrons in the source diffusion region that have enough energy to pass over the source-channel barrier and reach the drain side (see Figure 5.2). These electrons will create a nonzero I_d for $V_{ds} > 0$. This is known as the weak-inversion or subthreshold current, and it is the dominant leakage mechanism in modern devices. Since the number of these carriers is exponentially increased by an applied gate voltage below the threshold voltage, the weak-inversion current is represented by a straight line with a finite slope in a semilog plot as shown in Figure 5.1. The reverse of the slope of this line is known as the subthreshold swing S, and it has units of millivolts of the gate voltage per decade of the drain current. The ideal value of S is approximately 60 mV per decade. This value is a fundamental limit and represents the fact that V_g needs to be increased by at least 60 mV in order to achieve a factor of 10 increase in current over the potential barrier. To beat this limit, the carriers must additionally tunnel through the barrier. Tunneling may be afforded by a microelectromechanical system switch [1] or a tunneling transistor based on gate-induced band-to-band-tunneling [2].

Why is the subthreshold leakage a big concern in the recent CMOS technology nodes? As can be visualized through Figure 5.1, reducing V_t, which is equivalent to shifting the entire curve to the left, or increasing S, which makes the slope shallower, will increase the off-state leakage current exponentially, and hence will increase the static power dissipation. Short-channel MOSFETs inherently have smaller threshold voltages owing to two-dimensional electrostatics which originates from the proximity of the source and drain regions and their charge sharing with the gate (the effect known as V_t roll-off). The effect is enhanced at higher drain voltages in short-channel

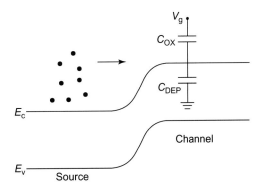

FIGURE 5.2

The potential barrier at the source/channel determines the subthreshold current. The current can be increased by 10 times for a 60 mV decrease in the potential barrier or equivalently for a $60 \times (1 + C_{DEP}/C_{OX})$ mV increase in V_g. The thin body of the FinFET becomes fully depleted for a small value of applied V_g; this makes C_{DEP} equal zero and S roughly 60 mV for the FinFET.

devices because the drain region is close enough that the drain voltage can affect and lower the source-channel barrier height at the channel-dielectric interface (the effect known as drain-induced barrier lowering, DIBL). Typical values of DIBL for high-performance, 32 nm node, planar bulk MOSFETs are around 100 mV/V; this means there is a 100 mV shift in V_t for 1 V of applied drain voltage. In very short channel devices, S is also affected and becomes larger (the effect known as subthreshold swing degradation). This is because the drain is so close that it can lower the source-channel barrier height for paths a few nanometers below the surface, resulting in subsurface leakage. Typical values of S for high-performance, 32 nm node, planar bulk MOSFETs are in the range of 70-100 mV per decade.

By providing a tighter electrostatic control around the channel, the FinFET has demonstrated a great ability in controlling short-channel effects and suppression of the off-state leakage current. The values of DIBL and S for a high-performance, 22 nm node FinFET are approximately 50 mV/V and approximately 70 mV per decade, respectively, leading to low values of I_{off} in the range of 20-100 nA/μm [3]. Still, determining the subthreshold behavior in scaled FinFETs through a careful modeling of effects which reduce V_t or degrade S (and hence worsen the off-state leakage current) is of great importance for IC designs. This is especially critical for low-power, mobile circuit applications. As discussed in Chapter 3, the BSIM-CMG model employs a current equation in its core model which is valid for a long-channel FinFET from weak inversion (subthreshold) to strong inversion. For the implementation of V_t roll-off, DIBL, and S degradation models, please refer to the real device models described in Chapter 4.

5.2 GATE-INDUCED SOURCE AND DRAIN LEAKAGES

Figure 5.3 illustrates the cross-section of an n-channel, double-gate FinFET and its energy-band diagram for the gate-drain overlap region when a low gate voltage and a high drain voltage are applied. If the band bending at the oxide interface is greater than or equal to the energy band gap E_g of the drain material, band-to-band tunneling will take place. The electrons in the valence band of the n-type drain will tunnel through the thinned band gap into the conduction band, and they will be collected at the drain contact to be a part of the drain current, whereas the reaming holes will be collected at the substrate contact (the source contact in the case of a FinFET on an SOI substrate) and will contribute to the substrate (source) leakage. This phenomenon, which was first elucidated and modeled by researchers at the University of California, Berkeley [4], discerns a potential major contributor to the off-state leakage current (see Figure 5.4) and is called the gate-induced drain leakage (GIDL) current. Depending on the voltages applied, there might also exist a gate-induced source leakage (GISL) current.

But what are the prerequisites for the GIDL current to flow? First, there must be band bending greater than E_g so that the valence band energy states overlap the conduction band energy states as shown in Figure 5.3. In that case the semiconductor

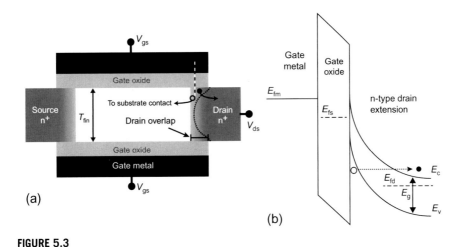

(a)

(b)

FIGURE 5.3

(a) Cross-section of the fin of a FinFET, and (b) illustration of the energy-band diagram along the dashed line in (a). The substrate contact is normal to and below the page.

surface in the gate-drain overlap region is in *deep depletion*, with the band bending being much larger than $2\varphi_B$.[3]

The surface potential can exceed $2\varphi_B$ because there is no inversion hole layer at the surface. There is no hole layer at the surface because any hole there would drift and diffuse to the body/substrate because of the built-in junction potential plus any substrate-drain reverse bias. However, with a forward-biased substrate-drain junction, the holes may remain at the interface and form an inversion layer and cause the band bending to be pinned at roughly $2\varphi_B$, a value smaller than E_g, thereby suppressing the GIDL current. In the case of a FinFET on an SOI substrate, holes build up in the floating body and raise the body potential until the body-source junction is slightly forward biased, enabling the GIDL-generated holes to be injected into the n$^+$ source. Second, the electric field needs to be large; that is, the tunneling barrier needs to be narrow. Compared with a planar MOSFET, both of these conditions are more difficult to meet in a FinFET because the potential at both sides of the thin fin is raised or lowered by the same V_g. Therefore, lightly doped and very thin fin FinFETs can have negligible GIDL. Try to convince yourself of this by looking at Figure 5.3 and remember the Poisson equation.

In addition, defects or traps in tunneling lead to trap-assisted band-to-band tunneling by providing stepping stones along the tunneling path; therefore, GIDL current is larger in the presence of defects created by ion implantation. Use of solid-source diffusion instead of implantation for drain creation or use of a laser for activation of dopants and annealing has been shown to reduce GIDL [5, Chapter 3].

---

[3] φ_B is the difference between the Fermi potential and the intrinsic potential in the drain.

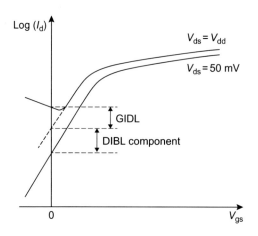

FIGURE 5.4

Contributions of DIBL and GIDL to the transistor's off-state leakage current. The position of the dip caused by GIDL will vary around $V_{gs} = 0$ depending on V_{dd}, the channel material, doping, and trap density.

5.2.1 GIDL/GISL CURRENT FORMULATION IN BSIM-CMG

The band-to-band tunneling current density from the Wentzel-Kramers-Brillouin (WKB) approximation is given by

$$J = A \times E_s \times e^{-B/E_s}, \tag{5.1}$$

where A is a preexponential constant related to the density of states of both the emitting side and the receiving side, B is a physical exponential parameter which depends on E_g and the carrier's effective mass in the tunneling direction (approximately 20 MV/cm for silicon), and E_s is the surface electric field in the drain. With use of Gauss's law at the onset of GIDL, when the band bending in the drain is equal to E_g, E_s is given by

$$E_s = \frac{V_{ds} - V_{gs} + V_{fbsd} - E_g}{\epsilon_{ratio} \times EOT}, \tag{5.2}$$

where V_{fbsd} is the flat-band voltage between the gate and the drain, ϵ_{ratio} is the ratio of the dielectric constant of the substrate material EPSRSUB over that of silicon dioxide, and EOT is the equivalent oxide thickness. Equations (5.1) and (5.2) lead to the following equation for the GIDL current in BSIM-CMG:

$$I_{gidl0} = NFIN_{total} \times W_{eff} \times AGIDL \times \left(\frac{V_{dg} + V_{fbsd} - EGIDL}{\epsilon_{ratio} \times EOT} \right)^{PGIDL}$$

$$\times e^{(-(\epsilon_{ratio} \times EOT \times BGIDL)/(V_{dg} + V_{fbsd} - EGIDL))} \tag{5.3}$$

where the constants A and B in Equation (5.1) and E_g in Equation (5.2) have been replaced by the model parameters AGIDL and BGIDL, and EGIDL, respectively, and PGIDL has been introduced for more flexibility in fitting the measured data.

The GISL current is calculated in the same manner:

$$I_{gisl0} = NFIN_{total} \times W_{eff} \times AGISL \times \left(\frac{V_{sg} + V_{fbsd} - EGISL}{\epsilon_{ratio} \times EOT} \right)^{PGISL}$$

$$\times e^{(-(\epsilon_{ratio} \times EOT \times BGISL)/(V_{sg} + V_{fbsd} - EGISL))} \tag{5.4}$$

In addition to the V_{dg} dependence present in Equation (5.3), in bulk FinFETs (BULKMOD $\neq$ 0), the GIDL current is also affected by the substrate bias for small values of V_{de} (the drain to substrate voltage) as the deep depletion condition in the drain surface starts to fail. The total GIDL current is obtained by multiplying I_{gidl0} from Equation (5.3) by an empirical factor for modeling the low V_{de} effect as follows:

$$I_{gidl} = \begin{cases} I_{gidl0} \times \dfrac{V_{de}^3}{CGIDL + V_{de}^3} & V_{de} \geq 0 \\ 0 & V_{de} < 0 \end{cases} \tag{5.5}$$

In Equation (5.5), CGIDL is a non-negative fitting parameter. A similar equation holds for the GISL current:

$$I_{gisl} = \begin{cases} I_{gisl0} \times \dfrac{V_{se}^3}{CGISL + V_{se}^3} & V_{se} > 0 \\ 0 & V_{se} \leq 0 \end{cases} \tag{5.6}$$

For a FinFET on an SOI substrate (BULKMOD = 0), I_{gidl0} and I_{gisl0} are multiplied by V_{ds} and V_{sd}, respectively. These terms are negligible in comparison with the exponential terms proceeding them, but will guarantee that no GIDL or GISL current is flowing when the drain and source are at the same voltage.

5.3 GATE OXIDE TUNNELING

For decades, to help the gate to keep its supremacy against the drain in controlling the source-to-channel barrier, the gate silicon dioxide (silicon oxynitride) thickness was scaled down in proportion to L_g. In the early years of this century, the tunneling through the scaled silicon oxynitride started to dominate the transistor's off-state leakage current, making it intolerable. A thicker dielectric layer with a higher dielectric constant (κ) was required. A thick, high-κ gate oxide could retain the control of the gate over the channel with orders of magnitude reduction in dielectric leakage current compared with SiO_2 of the same EOT. Furthermore, a metal gate eliminates the polysilicon gate depletion effect which was effectively increasing the gate dielectric thickness and thus reducing the gate control of the channel. High-κ oxides, in general, were also found to form a better interface with metal gates than the traditional polysilicon gate. This led to the introduction of high-κ metal-gate technology in the 45 nm node [6], which was extended to the successive nodes. However, the gate tunneling leakage through the gate oxide remains a significant and increasing concern as each new technology generation requires a smaller EOT.

5.3.1 GATE OXIDE TUNNELING FORMULATION IN BSIM-CMG

The gate oxide tunneling in the BSIM-CMG model inherits a similar formulation to that of BSIM4. Although the formulation has been derived for a polysilicon-silicon oxide gate stack, it turns out to be accurate enough to be used for high-κ metal-gate technology thanks to its flexibility. As illustrated in Figure 5.5, the gate tunneling current is composed of several mechanisms: the gate-to-body leakage current I_{gb}, the leakage currents through gate-to-source and gate-to-drain overlaps I_{gs} and I_{gd}, and the gate-to-inverted channel tunneling current I_{gc}. Part of I_{gc} is collected by the source (I_{gcs}), while the rest goes to the drain (I_{gcd}). I_{gb}, I_{gs}, I_{gd}, and I_{gc} are determined from the MOS capacitor, dielectric leakage model described below. Then, I_{gc} is extended to nonzero V_{ds} and partitioned into I_{gcs} and I_{gcd}.

On the basis of the early work of Lee and Hu [7], the dielectric tunneling leakage current density of a MOS capacitor can be modeled as

$$J_g = A \times \left(\frac{\text{TOXREF}}{\text{TOXG}}\right)^{\text{NTOX}} \times \frac{V_{ge} \times V_{aux}}{\text{TOXG}^2} \times e^{-B \times (\alpha - \beta.|V_{ox}|) \times (1 + \gamma.|V_{ox}|) \times \text{TOXG}}, \quad (5.7)$$

where $A = q^2/(8\pi h \varphi_b)$, $B = \left(8\pi \sqrt{2q m_{ox} \varphi_b^{3/2}}\right)/3h$, φ_b is the tunneling barrier height, m_{ox} is the effective carrier mass in the oxide, TOXG is the oxide thickness (different from the physical oxide thickness TOXP to introduce more flexibility), TOXREF is the reference oxide thickness at which all the parameters are extracted, NTOX is a fitting parameter that defaults to 1, V_{aux} is an auxiliary function which represents the density of tunneling carriers as well as available energy states to tunnel

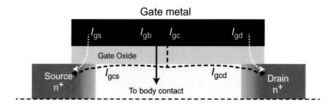

FIGURE 5.5

Half cross-section of the fin of the FinFET shown in Figure 5.2. The components of the tunneling current are shown. I_{gs} and I_{gd} are tunneling currents in the gate-to-source/drain overlap regions; I_{gb} flows between the gate and the body; I_{gc} is the gate-to-channel tunneling current and it is partitioned into I_{gcs} and I_{gcd}, which flow out of the source and drain, respectively.

into, and α, β, and γ are fitting parameters. Depending on the mode of operation (accumulation or depletion/inversion) and the gate tunneling component of interest, the values of m_{ox}, φ_b, and V_{aux} will be different, as explained below.

5.3.2 GATE-TO-BODY TUNNELING CURRENT IN DEPLETION/INVERSION

Figure 5.6 schematically demonstrates the dominant leakage mechanism between the gate and the body in depletion/inversion, represented by I_{gbinv}. In both p-type MOS (PMOS) and n-type MOS (NMOS), the electrons tunnel from the valence band of the body into the gate material. For this case, the values of A, B, and V_{aux} for a Si-SiO$_2$ interface (silicon as the body and silicon oxide as the gate oxide) are calculated to be

$$A = 3.75956 \times 10^{-7} \left(\frac{A}{V^2} \right), \tag{5.8}$$

$$B = 9.82222 \times 10^{11} \left(\frac{g}{Fs^2} \right)^{0.5}, \tag{5.9}$$

and

$$V_{aux,igbinv} = NIGBINV \times \frac{kT}{q} \times \ln \left(1 + e^{(V_{ox} - EIGBINV)/(NIGBINV \times kT/q)} \right). \tag{5.10}$$

In Equation (5.10), NIGBINV and EIGBINV are model parameters.

The total gate tunneling current I_{gbinv} is then given by

$$I_{gbinv} = NFIN_{total} \times W_{eff} \times L_{eff} \times A \times \left(\frac{TOXREF}{TOXG} \right)^{NTOX} \times \frac{V_{ge} \times V_{aux,igbinv}}{TOXG^2}$$
$$\times e^{-B \times (AIGBINV(T) - BIGBINV.q_{ia}) \times (1 + CIGBINV.q_{ia}) \times TOXG}, \tag{5.11}$$

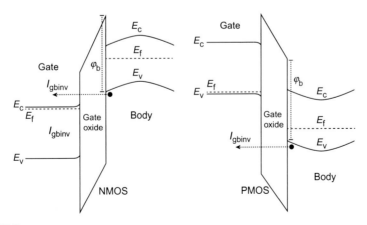

FIGURE 5.6

In both NMOS and PMOS, tunneling of valance-band electrons from the body into the gate is the principal cause of the gate-to-body tunneling current in inversion, I_{gbinv}.

where α, β, γ, and V_{ox} in Equation (5.7) have been replaced by the model parameters AIGBINV(T), BIGBINV, CIGBINV, and the average charge in the channel, q_{ia}, respectively.[4] The last approximation is valid since we assume that the fin is fully depleted and the body charge q_{ba} is a fixed value which can be incorporated into other model parameters.

5.3.3 GATE-TO-BODY TUNNELING CURRENT IN ACCUMULATION

In accumulation, the dominant leakage current between the gate and the body, I_{gbacc}, is the tunneling of conduction-band electrons. In NMOS, the electrons tunnel from the conduction band of the gate material into the conduction band of the body, and in PMOS they tunnel in the reverse direction (see Figure 5.7). For this case, the values of A, B, and V_{aux} for a polysilicon-silicon oxide-silicon structure are calculated to be

$$A = 4.97232 \times 10^{-7} \ \left(A/V^2\right),$$ (5.12)

$$B = 7.45669 \times 10^{11} \left(\frac{g}{Fs^2}\right)^{0.5},$$ (5.13)

and

$$V_{aux,igbinv} = \text{NIGBACC} \times \frac{kT}{q} \times \ln\left(1 + e^{(V_{fb}-V_{ge})/(\text{NIGBACC} \times kT/q)}\right).$$ (5.14)

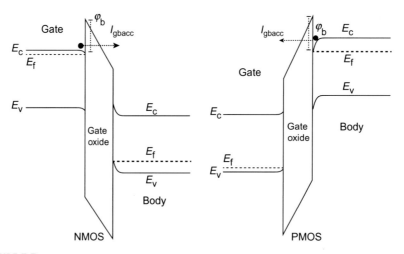

FIGURE 5.7

In accumulation, conduction-band electrons tunnel from the gate into the body in NMOS and from the body into the gate in PMOS.

---

[4]Note that all the charges in the BSIM-CMG model are normalized with respect to C_{ox}. That is why one can substitute a voltage with a charge.

In Equation (5.14), NIGBACC is a model parameter.

The total gate tunneling current I_{gbacc} is then given by

$$I_{gbacc} = NFIN_{total} \times W_{eff} \times L_{eff} \times A \times \left(\frac{TOXREF}{TOXG}\right)^{NTOX} \times \frac{V_{ge} \times V_{aux,igbinv}}{TOXG^2}$$

$$\times e^{-B \times (AIGBACC(T) - BIGBACC.q_{acc}) \times (1 + CIGBACC.q_{acc}) \times TOXG}, \tag{5.15}$$

where α, β, γ, and V_{ox} in Equation (5.7) have been replaced by the model parameters AIGBACC(T), BIGBACC, CIGBACC, and q_{acc}, respectively.

For BULKMOD $\neq 0$, I_{gb} (i.e., $I_{gbinv} + I_{gbacc}$) simply flows from the gate into the substrate. For BULKMOD = 0, I_{gb} mostly flows into the source because the potential barrier for holes is typically lower at the source side. To ensure continuity when V_{ds} switches sign, I_{gb} is partitioned into a source component I_{gbs} and a drain component I_{gbd} using the following partitioning scheme:

$$I_{gbs} = \left(I_{gbinv} + I_{gbacc}\right) \times W_f \tag{5.16}$$

$$I_{gbd} = \left(I_{gbinv} + I_{gbacc}\right) \times W_r, \tag{5.17}$$

where

$$W_f = \frac{1}{2} + \frac{1}{2} \times \tanh\left(\frac{0.6 \times q \times V_{ds}}{kT}\right) \tag{5.18}$$

and

$$W_r = \frac{1}{2} - \frac{1}{2} \times \tanh\left(\frac{0.6 \times q \times V_{ds}}{kT}\right). \tag{5.19}$$

5.3.4 GATE-TO-CHANNEL TUNNELING CURRENT IN INVERSION

As shown in Figure 5.8, in inversion, the electrons (holes in PMOS) tunnel from the inversion channel into the conduction band of the gate (valance band for PMOS). This results in different values of A and B for NMOS and PMOS:

$$A = \begin{cases} 4.97232 \times 10^{-7} \left(A/V^2\right) & \text{for NMOS,} \\ 3.42536 \times 10^{-7} \left(A/V^2\right) & \text{for PMOS} \end{cases} \tag{5.20}$$

and

$$B = \begin{cases} 7.45669 \times 10^{11} \left(\frac{g}{Fs^2}\right)^{0.5} & \text{for NMOS,} \\ 1.16645 \times 10^{12} \left(\frac{g}{Fs^2}\right)^{0.5} & \text{for PMOS.} \end{cases} \tag{5.21}$$

The auxiliary function $V_{aux,igc}$ can be shown to be

$$V_{aux,igc} = V_{ox}/V_{ge} \times \left(V_{ge} - 0.5 \cdot V_{dsx} + 0.5 \cdot V_{es} + 0.5 \cdot V_{ed}\right). \tag{5.22}$$

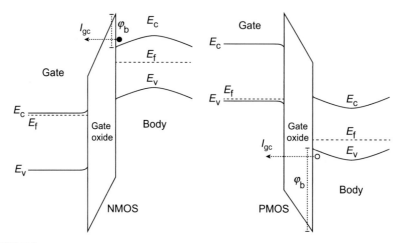

FIGURE 5.8

In inversion for NMOS, conduction-band electrons tunnel from the channel into the gate, whereas in PMOS, the valance-band holes tunnel from the channel into the gate.

The total gate-to-channel tunneling component at zero V_{ds} can be written as

$$I_{gc0} = \text{NFIN}_{\text{total}} \times W_{\text{eff}} \times L_{\text{eff}} \times A \times \left(\frac{\text{TOXREF}}{\text{TOXG}} \right)^{\text{NTOX}} \times \frac{V_{ge} \times V_{\text{aux,igcv}}}{\text{TOXG}^2}$$

$$\times e^{-B \times (\text{AIGC}(T) - \text{BIGBC}.q_{ia}) \times (1 + \text{CIGC}.q_{ia}) \times \text{TOXG}}. \tag{5.23}$$

To consider the drain bias effect, a current continuity equation is solved analytically along the channel which extends I_{gc0} to nonzero V_{ds} and splits it into two components, I_{gcs} and I_{gcd}. For a detailed discussion on the derivation of this physical current partitioning factor, refer to [8]. The expressions for I_{gcs} and I_{gcd} are as follows:

$$I_{gcs} = I_{gc0} \frac{\text{PIGCD} \times |V_{\text{dseff}}| + e^{(-\text{PIGCD}.V_{\text{dseff}})} - 1}{\text{PIGCD}^2 \times V_{\text{dseff}}^2}, \tag{5.24}$$

$$I_{gcd} = I_{gc0} \frac{(\text{PIGCD} \times |V_{\text{dseff}}| + 1) \times e^{(-\text{PIGCD}.V_{\text{dseff}})}}{\text{PIGCD}^2 \times V_{\text{dseff}}^2}, \tag{5.25}$$

where PIGCD is a fitting parameter added for flexibility with a default value of unity.

5.3.5 GATE-TO-SOURCE/DRAIN TUNNELING CURRENT

The n^+ (p^+) gate to n^+ (p^+) source and drain currents I_{gs} and I_{gd} are principally caused by the tunneling of the conduction-band electrons in NMOS and valance-band holes in PMOS as shown in Figure 5.9. In NMOS, the electrons tunnel from the conduction band of the body into the gate. In PMOS, the holes tunnel from the valance band of the body into the gate.

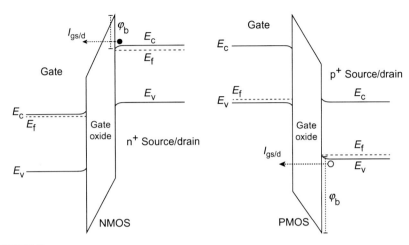

FIGURE 5.9

In NMOS the tunneling of the conduction-band electrons and in PMOS the tunneling of the valance-band holes make the source/drain-gate overlap leakage. The figure shows the band diagrams for an inverted channel. In accumulation, the direction of the tunneling is reversed.

For this case, the parameters A and B are naturally equal to those given by Equation (5.20) and Equation (5.21), respectively. If the gate material is a metal, V_{aux} is also simplified to be equal to $|V_{gs}|$ and $|V_{gd}|$ for I_{gs} and I_{gd}, respectively.

The total gate-to-source extension tunneling component is

$$I_{gs} = \text{NFIN}_{total} \times W_{eff} \times \text{DLCIGS} \times A \times \left(\frac{\text{TOXREF}}{\text{TOXG} \times \text{POXEDGE}}\right)^{\text{NTOX}}$$

$$\times \frac{V_{gs} \times |V_{gs}|}{(\text{TOXG} \times \text{POXEDGE})^2} \times e^{-B \times (\text{AIGS}(T) - \text{BIGS}.|V_{gs}|) \times (1 + \text{CIGS}.|V_{gs}|) \times \text{TOXG} \times \text{POXEDGE}}.$$

$$(5.26)$$

In Equation (5.26), DLCIGS is the length of the gate-source overlap region and POXEDGE is a factor for the gate oxide thickness in the source/drain extension regions.

Similarly, for the total gate-to-drain extension tunneling component, we have

$$I_{gd} = \text{NFIN}_{total} \times W_{eff} \times \text{DLCIGD} \times A \times \left(\frac{\text{TOXREF}}{\text{TOXG} \times \text{POXEDGE}}\right)^{\text{NTOX}}$$

$$\times \frac{V_{gd} \times |V_{gd}|}{(\text{TOXG} \times \text{POXEDGE})^2}$$

$$\times e^{-B \times (\text{AIGD}(T) - \text{BIGD} \times |V_{gd}|) \times (1 + \text{CIGD} \times |V_{gd}|) \times \text{TOXG} \times \text{POXEDGE}}, \qquad (5.27)$$

where DLCIGD is the length of the gate-drain overlap region.

5.4 IMPACT IONIZATION

In the transistor on state, because of the high electric field near the drain end of the channel, carriers in this region can gain enough kinetic energy to ionize the lattice atoms when they collide. This collision frees an electron from the valance-band and leaves a hole behind. The generated hole will drift to the substrate and it will increase the substrate leakage. The released high-energy electron (hot carrier) is collected by the drain and it will be a part of the drain current. Also, there is chance that the generated hot electron travels along the gate field and penetrates into the gate oxide. Hot carrier injection into the gate oxide over time can damage the oxide and cause reliability problems.

The local impact ionization current $I_{ii}(y)$ can be written as a function of the channel current (increase in the number of carriers will increase the chance of collisions) and the strength of the local electric field (the stronger the electric field, the higher the kinetic energy of the carriers) as follows:

$$I_{ii}(y) = I_{ds}.A_i e^{-B_i/E_l(y)} \tag{5.28}$$

where A_i and B_i are two material constants and represent how often the impact ionization events take place and the critical field to trigger the events, respectively, and $E_l(y)$ is the longitudinal electric field along the transport direction. By integrating Equation (5.28) along the length of the channel where velocity saturation happens, we can write the total impact ionization current as

$$I_{ii} = I_{ds}.A_i \int_{y=0}^{y=\hat{l}} e^{-B_i/E_l(y)} dy, \tag{5.29}$$

where $y = 0$ is the starting point of the velocity saturation region and $\hat{l}$ is the length of this region. The integration in Equation (5.29) can be performed (see [9, Chapter 4] for details) to give the following equation for the BSIM-CMG impact ionization model:

$$I_{ii} = \frac{A_i}{B_i}.I_{ds}.(V_{ds} - V_{dsat}).e^{(-B_i.\lambda)/(V_{ds}-V_{dsat})}, \tag{5.30}$$

where V_{dsat} is the saturation voltage and λ is the characteristic length (see Chapter 4). The first impact ionization model (IIMOD = 1) implements Equation (5.30) as

$$I_{ii} = \left(ALPHA1 + \frac{ALPHA0}{L_{eff}}\right).I_{ds}.(V_{ds} - V_{dseff}).e^{-BETA0/(V_{ds}-V_{dseff})}. \tag{5.31}$$

In Equation (5.31), V_{dseff} is the effective drain voltage resulting from the smooth transition of V_{ds} to V_{dsat} (see Chapter 4), ALPHA1 and BETA0 are a fitting parameter, and the term ALPHA0/L_{eff} has been introduced to improve the length dependence of I_{ii} over a wide range of channel lengths.

There are approximations involved in deriving Equations (5.30) and (5.31), including a linear dependence of $E_l(y)$ on $(V_{ds} - V_{dsat})$. The BSIM-CMG's second impact ionization model can be activated (IIMOD = 2) and used if a more flexible model is needed:

$$I_{ii} = \left(ALPHA1 + \frac{ALPHA0}{L_{eff}} \right).I_{ds}.\exp\left(\frac{V_{diff}}{BETAII2 + BETAII1.V_{diff} + BETAII0.V_{diff}^2} \right),$$
(5.32)

$$V_{diff} = V_{ds} - V_{dsatii},$$
(5.33)

$$V_{dsatii} = V_{gsStep}\left(1 - \frac{LII}{L_{eff}} \right),$$
(5.34)

$$V_{gsStep} = \left(\frac{ESATII.L_{eff}}{1 + ESATII.L_{eff}} \right)\left(\frac{1}{1 + SII1.V_{gsfbeff}} + SII2 \right)\left(\frac{SII0.V_{gsfbeff}}{1 + SIID.V_{ds}} \right).$$
(5.35)

Here, BETAII0, BETAII1, BETAII2, and SIID are parameters for V_{ds}-dependence, LII is a channel-length-dependent parameter, SII0, SII1, and SII2 are V_{gs}-dependent fitting parameters, and ESATII is the channel saturation field with the default value of 1×10^7 V/m.

REFERENCES

[1] H. Kam, V. Pott, R. Nathanael, J. Jeon, E. Alon, T.-J.K. Liu, Design and reliability of a micro-relay technology for zero-standby-power digital logic applications, Electron Devices Meeting (IEDM), 2009 IEEE International, 7–9 December 2009, pp. 1–4.
[2] K. Jeon, W.Y. Loh, P. Patel, et al., Si tunnel transistors with a novel silicided source and 46 mV/dec swing, in: VLSI Symp. Tech. Dig., June 2010, pp. 121–122.
[3] C.-H. Jan, U. Bhattacharya, R. Brain, S.-J. Choi, G. Curello, G. Gupta, W. Hafez, M. Jang, M. Kang, K. Komeyli, T. Leo, N. Nidhi, L. Pan, J. Park, K. Phoa, A. Rahman, C. Staus, H. Tashiro, C. Tsai, P. Vandervoorn, L. Yang, J.-Y. Yeh, P. Bai, A 22 nm SoC platform technology featuring 3-D tri-gate and high-k/metal gate, optimized for ultra low power, high performance and high density SoC applications, Electron Devices Meeting (IEDM), 2012 IEEE International, 10–13 December 2012, pp. 3.1.1–3.1.4.
[4] T.-Y. Chan, J. Chen, P.-K. Ko, C. Hu, The impact of gate-induced drain leakage current on MOSFET scaling, Electron Devices Meeting, 1987 International, vol. 33, 1987, pp. 718–721.
[5] C.C. Hu, Modern Semiconductor Devices for Integrated Circuits, Prentice Hall, Upper Saddle River, 2009.
[6] C. Auth, M. Buehler, A. Cappellani, C.-H. Choi, G. Ding, W. Han, S. Joshi, B. McIntyre, M. Prince, P. Ranade, J. Sandford, C. Thomas, 45 nm high-k + metal-gate strain-enhanced transistors, Intel Technol. J. 12 (2) (2008) 77–86.
[7] W.-C. Lee, C. Hu, Modeling gate and substrate currents due to conduction- and valence-band electron and hole tunneling [CMOS technology], 2000 Symposium on VLSI Technology. Digest of Technical Papers, 13–15 June 2000, pp. 198–199.

[8] K.M. Cao, W.-C. Lee, W. Liu, X. Jin, P. Su, S.K.H. Fung, J.X. An, B. Yu, C. Hu, BSIM4 gate leakage model including source-drain partition, Electron Devices Meeting, 2000. IEDM'00. Technical Digest International, 10–13 December 2000, pp. 815–818.

[9] W. Liu, C.C. Hu, BSIM4 and MOSFET Modeling for IC Simulation, World Scientific, Singapore, 2011.

Charge, capacitance, and non-quasi-static effects

CHAPTER OUTLINE

6.1 Terminal charges 144
      6.1.1 Gate charge... 144
      6.1.2 Drain charge.. 145
      6.1.3 Source charge 146
6.2 Transcapacitances 146
6.3 Non-quasi-static effects models 147
      6.3.1 Relaxation time approximation model 149
      6.3.2 Channel-induced gate resistance model 151
      6.3.3 Charge segmentation model 152
References 155

Essential to IC design with a FinFET is a compact model that can accurately predict the device dynamic behavior in addition to DC/operating point analysis. This is because in practical circuits, the transistor's terminal voltages are not static and they vary with the applied, time-dependent, large or small signals. The functionality of such circuits is determined through AC and transient analyses. AC analysis needs the capacitances associated with the terminals of the transistor. To analyze the transient behavior, it is essential to model the charges stored in the transistor. The BSIM-CMG C-V model defines both the charges and the associated capacitances for the FinFET. This chapter discusses only the intrinsic transistor charge and capacitances, and the details of source/drain and gate overlap regions and real source and drain contacts are left for Chapter 7. Those will create parasitic overlap and fringe capacitances that should be added to the intrinsic model described here. This method is consistent with the historical approach toward transistor modeling, such as with traditional MOSFETs [1] and bipolar transistors [2], where electrostatics and transport within an internal transistor are used to create a core model, and external parasitic components (which will vary with the overall device structure) are then added around this core.

Advanced RF electronics also requires that compact models accurately predict the device high-frequency behavior at least up to the transistor's current-gain cutoff frequency (f_T), and preferably, somewhat beyond. This requires the incorporation of the non-quasi-static (NQS) effects in the compact model. We start this chapter by deriving the terminal charges and transcapacitances, and conclude by describing the BSIM-CMG's NQS models.

6.1 TERMINAL CHARGES
6.1.1 GATE CHARGE

The total gate charge Q_g is the integral of the local channel charge density $\acute{Q}_{ch}$[1] over the effective channel length L:

$$Q_g = -W \int_0^L \acute{Q}_{ch}(y)\, dy \tag{6.1}$$

where W is the total width of the FinFET and y is the transport direction along the channel. The value of $\acute{Q}_{ch}(y)$ from Gauss's law is given by

$$\acute{Q}_{ch}(y) = -C_{ox}\left[V_{gs} - V_{fb} - \psi(y)\right] \tag{6.2}$$

where C_{ox} is the oxide capacitance per unit area and ψ is the surface potential. To be able to evaluate the integral in Equation (6.1) analytically, a closed-form expression for the variation of ψ along the channel is desired. We start with the current continuity

$$I_{ds}(L) = I_{ds}(y) \quad \forall \ 0 \le y \le L \tag{6.3}$$

and the following simplified—but adequately accurate—expression for the current:

$$I_{ds}(y) = \frac{\mu W}{L}\left[h\left(\acute{Q}_{invs}\right) - h\left(\acute{Q}_{inv}(y)\right)\right]. \tag{6.4}$$

In Equation (6.4), $h(Q) = \frac{Q^2}{2C_{ox}} + 2V_{tm}Q$, V_{tm} is the thermal voltage given by $k_B T/q$, where k_B and T are the Boltzmann constant and the temperature, respectively, $\acute{Q}_{invs}$ is the inversion charge density at the source end, and $\acute{Q}_{inv}(y)$ is the inversion charge density at position y given by

$$\acute{Q}_{inv}(y) = -C_{ox}\left(V_{gs} - V_{fb} - \psi(y) - \frac{\acute{Q}_{bulk}}{C_{ox}}\right) \tag{6.5}$$

where $\acute{Q}_{bulk}$ is the fixed depletion charge density[2] and is given by $q \times$ NBODY $\times$ TFIN. Here NBODY is the doping concentration of the channel and TFIN is thickness of the channel (see Appendix for description of parameters).

---

[1] Throughout this chapter, charge quantities with a prime are charge density per unit area.
[2] The thin body of the FinFET becomes fully depleted for a small value of applied V_{gs}.

Using Equation (6.4), we can rewrite the equality in Equation (6.3) to read

$$\frac{h\left(\acute{Q}_{\text{invs}}\right) - h\left(\acute{Q}_{\text{invd}}\right)}{L} = \frac{h\left(\acute{Q}_{\text{invs}}\right) - h\left(\acute{Q}_{\text{inv}}\,(y)\right)}{y} \tag{6.6}$$

Substituting the values of $\acute{Q}_{\text{invs}}$ and $\acute{Q}_{\text{invd}}$ obtained from Equation (6.5) into Equation (6.6), we can write

$$\frac{(B - \psi_s - \psi_d)\,(\psi_d - \psi_s)}{L} = \frac{(B - \psi_s - \psi)\,(\psi - \psi_s)}{y}. \tag{6.7}$$

where $B = 2\left(V_{\text{gs}} - V_{\text{fb}} - \acute{Q}_{\text{bulk}}C_{\text{ox}} + 2V_{\text{tm}}\right)$, and ψ_s and ψ_d are the source-end and drain-end surface potentials, respectively. Equation (6.7) can be solved to get $\psi(y)$; however, as will be seen shortly, it would be more efficient to solve it in terms of y

$$y = \frac{L\,(B - 2\psi)\,(\psi - \psi_s)}{(B - \psi_s - \psi_d)\,(\psi_d - \psi_s)} \tag{6.8}$$

and differentiate it to get

$$dy = \frac{L\,(B - 2\psi)\,d\psi}{(B - \psi_s - \psi_d)\,(\psi_d - \psi_s)}. \tag{6.9}$$

Finally, using Equations (6.2) and (6.9) in Equation (6.1) results in the following expressions for the gate charge as a function of the gate voltage and the source-end and drain-end surface potentials:

$$\frac{Q_g}{WC_{\text{ox}}} = \int_0^L \left(V_{\text{gs}} - V_{\text{fb}}\right) dy - \int_0^L \psi\,dy = \left(V_{\text{gs}} - V_{\text{fb}}\right)L - \int_{\psi_s}^{\psi_d} \frac{\psi L\,(B - 2\psi)}{(B - \psi_s - \psi_d)\,(\psi_d - \psi_s)}\,d\psi, \tag{6.10}$$

$$\frac{Q_g}{WLC_{\text{ox}}} = V_{\text{gs}} - V_{\text{fb}} - \int_{\psi_s}^{\psi_d} \frac{\psi\,(B - 2\psi)}{(B - \psi_s - \psi_d)\,\psi_{\text{ds}}}\,d\psi$$

$$= V_{\text{gs}} - V_{\text{fb}} - \frac{(\psi_s + \psi_d)}{2} + \frac{(\psi_d - \psi_s)^2}{6\,(B - \psi_s - \psi_d)}. \tag{6.11}$$

6.1.2 DRAIN CHARGE

The sum of the drain charge and the source charge should equal the total channel inversion charge. The task of dividing the channel inversion charge between the source and the drain is called charge partitioning. A simplistic partitioning scheme is 50:50 partitioning, which divides the channel inversion charge equally between the source and the drain. This is valid only if the device is symmetric and the drain-source voltage V_{ds} is low. As V_{ds} increases above zero, this scheme becomes more and more erroneous. Another arbitrary assignment of charge is 0:100 partitioning, which allots all the channel inversion charge to the source, and the drain charge is set to zero. However, a commonly used and physically valid partitioning scheme is 40:60

partitioning, also known as Ward-Dutton partitioning [3]. According to this scheme, the drain charge is given by

$$Q_d = -WC_{ox} \int_0^L \frac{y}{L} \left(V_{gs} - V_{fb} - \psi - \frac{\acute{Q}_{bulk}}{C_{ox}} \right) dy. \tag{6.12}$$

Substituting dy from Equation (6.9) results in

$$Q_d = -WC_{ox} \left(\frac{V_{gs} - V_{fb} - \frac{\acute{Q}_{bulk}}{C_{ox}}}{2} L - \int_{\psi_s}^{\psi_d} \psi y \frac{(B - 2\psi)}{(B - \psi_s - \psi_d)(\psi_d - \psi_s)} d\psi \right). \tag{6.13}$$

Furthermore, substituting y from Equation (6.8), we can write

$$\frac{-Q_d}{WLC_{ox}} = \frac{V_{gs} - V_{fb} - \acute{Q}_{bulk}/C_{ox}}{2} - \int_{\psi_s}^{\psi_d} \frac{\psi (B - 2\psi)(\psi - \psi_s)(B - \psi_s - \psi_d)}{(B - \psi_d - \psi_s)^2 (\psi_d - \psi_s)^2} d\psi. \tag{6.14}$$

The integral on the right-hand side of Equation (6.14) can be evaluated to ultimately give

$$\frac{-Q_d}{WLC_{ox}} = \frac{V_{gs} - V_{fb} - \acute{Q}_{bulk}/C_{ox}}{2} - \frac{\psi_s + \psi_d}{4} + \frac{(\psi_d - \psi_s)^2}{60 (B - \psi_s - \psi_d)}$$
$$+ \frac{(5B - 6\psi_s - 4\psi_d)(B - 2\psi_d)(\psi_s - \psi_d)}{60(B - \psi_s - \psi_d)^2}. \tag{6.15}$$

6.1.3 SOURCE CHARGE

On the basis of charge conservation, we can write

$$Q_s = - (Q_g + Q_d + Q_{bulk}) \tag{6.16}$$

where Q_{bulk} is the total bulk charge and is given by

$$Q_{bulk} = -W.L. \acute{Q}_{bulk}. \tag{6.17}$$

6.2 TRANSCAPACITANCES

All transcapacitances are derived from the terminal charges to ensure charge conservation. The transcapacitances are calculated as

$$C_{ij} = \begin{cases} -\dfrac{\partial Q_i}{\partial V_j}, & i \neq j, \\ \dfrac{\partial Q_i}{\partial V_j}, & i = j, \end{cases} \tag{6.18}$$

where Q_i and V_j are the charge and voltage associated with terminals i and j, respectively, while the voltages on the other terminals are held constant. In the BSIM-CMG model, the transcapacitances are calculated using the ddx() operator in Verilog-A, and they are reported as parts of the operating-point information. This operator

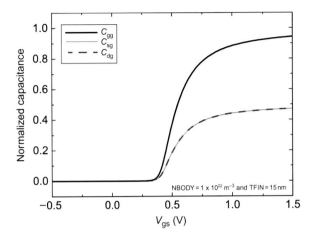

FIGURE 6.1

The transcapacitances C_{gg}, C_{sg}, and C_{dg} of an NMOS FinFET as a function of the gate to source voltage V_{gs} and for $V_{ds} = 0$. The body doping is 1×10^{22} m^{-3} and the fin thickness is 15 nm. The values have been normalized with respect to WLC_{ox}.

eliminates the need to write analytical expressions for the derivatives which might contain errors introduced during the derivation or implementation.

Figure 6.1 shows the transcapacitances C_{gg}, C_{sg}, and C_{dg} of an n-type MOS (NMOS) FinFET as a function of V_{gs} and for $V_{ds} = 0$. The transcapacitances start to increase as the device operating mode enters weak/moderate inversion. The C-V model partitions the channel charge equally between the source and the drain, and at strong inversion C_{sg} and C_{dg} reach $C_{gg}/2$. This is expected as for $V_{ds} = 0$ the source and the drain are electrically indistinguishable. However, for $V_{ds} = 1.2$ V, the Ward-Dutton partitioning scheme used in the model results in a 40/60 charge partitioning, and hence $C_{sg} \approx 0.6\, C_{gg}$ and $C_{dg} \approx 0.4\, C_{gg}$ (see Figure 6.2).

Figure 6.3 shows the transcapacitances C_{gg}, C_{sg}, C_{dg}, C_{gs}, and C_{gd} of an NMOS FinFET as a function of V_{ds} and for $V_{gs} = 1.2$ V. For $V_{ds} = 0$, the source and drain terminals are electrically indistinguishable; therefore, $C_{sg} = C_{dg}$ and $C_{gs} = C_{gd}$. Furthermore, noting that $C_{sg} + C_{dg} = C_{gs} + C_{gd} = C_{gg}$, we have $C_{sg} = C_{dg} = C_{gs} = C_{gd} = C_{gg}/2$. As the drain voltage increases, $C_{sg} \approx 0.6\, C_{gg}$ and $C_{dg} \approx 0.4\, C_{gg}$. In saturation (i.e., $V_{ds} > V_{dsat}$) $C_{gd} \to 0$ ($C_{gs} \to C_{gg}$) because any extra drain voltage is dropped in the pinch-off region and has no effect on the channel inversion charge, or in other words, the drain is getting discounted from the channel.

6.3 **NON-QUASI-STATIC EFFECTS MODELS**

The operation of a MOSFET would be defined by a self-consistent solution of Poisson's equation which describes the electrostatics and the current-continuity equation which governs the dynamics:

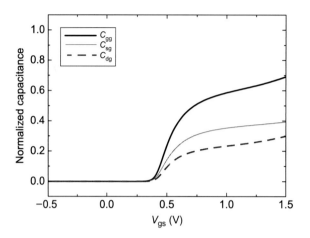

FIGURE 6.2

The transcapacitances from Figure 6.1 for $V_{ds} = 1.2$ V. The Ward-Dutton partitioning used in the model results in $C_{sg} \approx 0.6\ C_{gg}$ and $C_{dg} \approx 0.4\ C_{gg}$.

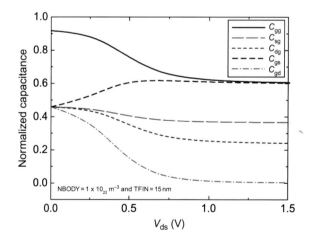

FIGURE 6.3

The transcapacitances C_{gg}, C_{sg}, C_{dg}, C_{gs}, and C_{gd} of an NMOS FinFET as a function of the drain to source voltage V_{ds} and for $V_{gs} = 1.2$ V. The BSIM-CMG capacitance-voltage model captures all the important physics as described in the text.

$$\frac{\partial^2 \psi}{\partial y^2} = \frac{\rho}{\epsilon_{ch}} \tag{6.19}$$

$$W\frac{\partial \acute{Q}(y,t)}{\partial t} = \frac{\partial I(y,t)}{\partial y}. \tag{6.20}$$

Here, y is the direction along the channel, ψ is the surface potential, ρ is the volume charge density (both mobile and fixed charges, in reciprocal cubic centimeters), $\acute{Q}$ is the channel mobile charge (in reciprocal square centimeters), t is time, and $I(y,t)$ is the current in the channel. The gradual channel approximation is assumed in the one-dimensional Poisson's equation; that is, the vertical field dominates over the lateral field along the channel. The continuity equation implies that there will be no buildup of charge at any position along the channel.

In modern MOSFET compact models such as BSIM-CMG, the quasi-static (QS) assumption is evoked (i.e., $\partial \acute{Q}/\partial t = 0$) and a steady-state drain current expression is derived to describe the DC operation. The AC or the small-signal operation is described by the terminal charges. The QS assumption implies that the steady-state current and the channel charges are established instantaneously after terminal voltages are applied to the device. This is not true when the device is subject to high slew rate (high dV/dt) signals—either large voltage swings in a short time or very high frequency signals with frequencies approaching or higher than the cutoff frequency f_T of the transistor. There is an inherent delay in the response of the transistor's currents and charges to applied voltages (often visualized as a distributed resistance-capacitance network). As research opens new avenues for circuit applications, today's circuit designs are either approaching f_T (such as terahertz CMOS) or being subject to high slew rate signals such as in CMOS RF power amplifiers. One needs to note that f_T is a function of the terminal voltages, especially the gate voltage (see Figure 2.12). Today, many circuit applications are subjected to near subthreshold gate voltages to lower the operating power, and hence they will inherently have lower f_T than that reported as a figure of merit for that device (which is usually at the highest operating voltage). For these cases, compact models are required to support the NQS mode of operation to be able to predict the circuit's behavior accurately.

The BSIM-CMG model offers three different NQS models. Each of these can be turned on/off by NQSMOD switch. Setting NQSMOD = 0 turns off all the NQS models and switches to QS calculations.

6.3.1 RELAXATION TIME APPROXIMATION MODEL

A simple and elegant way to capture the NQS behavior of the channel is to use the relaxation time approach and track the deficient or surplus charge in the channel, which is given by [4]

$$\frac{dQ_{def}}{dt} = \frac{dQ_{ch,eq}}{dt} - \frac{dQ_{def}}{\tau} \tag{6.21}$$

where $Q_{\text{def}} = Q_{\text{ch,nqs}} - Q_{\text{ch,eq}}$ is the deficient/surplus channel charge, $Q_{\text{ch,nqs}}$ is the channel charge considering NQS effects, $Q_{\text{ch,eq}}$ is the channel charge at steady-state equilibrium or the QS charge, and τ is the relaxation time constant given as follows:

$$\tau = \left[\text{XRCRG1} \cdot \frac{\text{NF}}{\text{NFIN}} \cdot \frac{\mu_{\text{eff}} \cdot W}{L} C_{\text{ox}} \left(q_{\text{ia}} + \text{XRCRG2} \frac{kT}{q} \right) \right]^{-1}. \tag{6.22}$$

In Equation (6.22), q_{ia} is the average charge in the channel and model parameters XRCRG1 and XRCRG2 are used to improve the model's flexibility. NF and NFIN denote the number of gate fingers and number of fins, respectively (see Appendix for description of parameters).

Within a SPICE environment, Equation (6.21) is implemented as a subcircuit whose node voltage tracks the deficient/surplus charge, Q_{def} (see Figure 6.4). The source and drain terminal charges are then given by

$$Q_{\text{d,nqs}} = X_{\text{part}} \frac{Q_{\text{def}}}{\tau} \tag{6.23}$$

and

$$Q_{\text{s,nqs}} = \left(1 - X_{\text{part}}\right) \frac{Q_{\text{def}}}{\tau} \tag{6.24}$$

respectively. In Equations (6.23) and (6.24), X_{part} is a bias-dependent partitioning fraction and it is approximated to its QS equivalent given by

$$X_{\text{part}} = \frac{Q_{\text{d}}}{Q_{\text{d}} + Q_{\text{s}}}. \tag{6.25}$$

The calculated terminal charges $Q_{\text{d,nqs}}$ and $Q_{\text{s,nqs}}$ replace their QS equivalents Q_{d} and Q_{s}, respectively. The NQS body charge $Q_{\text{bulk,nqs}}$ is the same as Q_{bulk} since the body terminal charge does not experience any NQS effect (e.g., the holes in an NMOS traverse through a p-doped bulk region quickly). A more rigorous evaluation of this approach found it to be accurate up to $2 \times f_{\text{T}}$ when compared with technology computer-aided design-based simulations [5]. This NQS model can be activated by setting NQSMOD = 2.

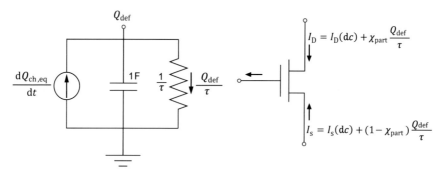

FIGURE 6.4

A relaxation time approach for capturing non-quasi-static effects. The resistance-capacitance subcircuit represents Equation (6.21) for SPICE implementation and it is solved self-consistently with the MOSFET.

Before using the model, one has to extract the parameters XRCRG1 and XR-CRG1 from the high-frequency measured data. The default values of 12 and 1 for XRCRG1 and XRCRG2, respectively, are obtained from the equivalent resistance of a distributed transmission line model with double-side contact (here source and drain).

6.3.2 CHANNEL-INDUCED GATE RESISTANCE MODEL

Another first-order model that helps to capture the NQS effects in the channel is the channel-induced gate resistance model [6]. This model is activated by setting NQSMOD = 1. In this method, the bias-dependent gate resistance R_{ii} is added at the gate terminal[3] whose value is proportional to the channel resistance (see Figure 6.5). This resistance should not to be confused with the physical gate electrode resistance. The induced gate resistance is obtained from the relaxation time constant in Equation (6.22) as follows [7, Appendix D]:

$$R_{ii} = \frac{\tau}{WLC_{ox}}. \tag{6.26}$$

While this method performs similarly to the relaxation time approach in terms of accuracy up to f_T, it is restricted in terms of applicability. This model should be used in cases where the gate terminal is excited only by an input signal. NQS effects for applications such as passive mixers or common-gate low-noise amplifiers where the

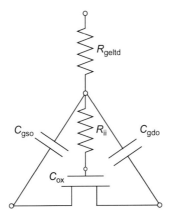

FIGURE 6.5

A channel-induced gate resistance R_{ii} is added in series with the gate physical electrode resistance R_{geltd} to capture the non-quasi-static effects for a gate-terminal-excited FinFET. C_{gso} and C_{gdo} are the parasitic capacitances.

---

[3] A gate node is introduced between the intrinsic gate and the physical gate electrode for this purpose. This node collapses to the intrinsic gate if the user turns off this model.

source terminal is excited cannot be captured by this model. However, the relaxation time constant approach is valid for all terminal excitations as we recalculate all the terminal charges.

6.3.3 CHARGE SEGMENTATION MODEL

If the channel length approaches infinitesimally small values, the carrier transit times through the channel tend to become small. For this device, the QS approximation is still valid. Using this concept, we can then visualize a transistor as a series of connected shorter channel length transistors or as we shall refer to them here as "charge segments" (owing to each smaller transistor carrying part of the whole channel charge). Figure 6.6 schematically shows such a representation. A charge-based version of the continuity equation in Equation (6.20) can be written as follows [7, Appendix D]:

$$\frac{\partial q}{\partial t} + \mu_{\text{eff}} V_t \frac{\partial}{\partial y} \left(\frac{(2q+1)}{0.5 \left(1 + \sqrt{1 + 2\left(\frac{\mu_{\text{eff}}}{v_{\text{sat}}} \frac{\partial q}{\partial y} \right)^2} \right)} \frac{\partial q}{\partial y} \right) = 0. \tag{6.27}$$

This equation is valid at any point along the channel and includes velocity saturation effects. The solution to this continuity equation along the channel captures the NQS effects. One way to accomplish this in a SPICE simulator environment is through a simple series connection of QS transistors, commonly known as the channel segmentation approach [8, 9]. If implemented within a compact model skillfully with care by not including short-channel effects for each segment (whose channel length is L/N_{SEG}, where N_{SEG} is the number of segments) or by not adding series resistance and parasitic capacitance multiple times for each segment, this method captures all the important features of NQS effects. However, often this approach is ill constrained, leading to longer simulation times or non-convergence.

A solution to the above continuity equation where the continuity constraint is imposed not just on the first derivative but even on the second and third derivatives is required. One such solution was developed using a spline-collocation method for

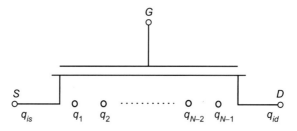

FIGURE 6.6

$N_{\text{SEG}} = N$ charge segments in a MOSFET channel for simulation of non-quasi-static effects. q_i represents the channel charge at the ith intermediate node.

a surface-potential-based MOSFET model in [10], which we will adopt for a charge-based model here.

Starting from Equation (6.27), we can further simplify it into the following form:

$$\frac{\partial q}{\partial t} + f\left(q, \frac{\partial q}{\partial y}, \frac{\partial^2 q}{\partial y^2}\right) = 0$$

$$f\left(q, \frac{\partial q}{\partial y}, \frac{\partial^2 q}{\partial y^2}\right) = \frac{\mu_{\text{eff}} V_t}{D_v}\left[2\left(\frac{\partial q}{\partial y}\right)^2 - \frac{2q+1}{D_v}\left(\frac{\mu_{\text{eff}}}{v_{\text{sat}}}\right)^2\left(\frac{\partial q}{\partial y}\right)^2\frac{\partial^2 q}{\partial y^2}\right],$$

$$D_v = \sqrt{1 + 2\left(\frac{\mu_{\text{eff}}}{v_{\text{sat}}}\right)^2\left(\frac{\partial q}{\partial y}\right)^2}. \tag{6.28}$$

In order to solve this complex partial differential equation in a SPICE environment, we need to convert it into a set of ordinary differential equations. In [10], the first-order weighted residuals method was extended to ensure current continuity up to the third order. Following a similar approach and assuming the channel is broken down into N_{SEG} segments, we can assume the charge in each segment is expressed by a cubic equation. For example, the inversion charge in the nth segment is given by

$$q(y) = a_n y^3 + b_n y^2 + c_n y + d_n \quad \frac{n-1}{N_{\text{SEG}}}L < y < \frac{n}{N_{\text{SEG}}}L. \tag{6.29}$$

The boundary conditions for the charge at the source and drain ends are their respective QS solutions; that is, $q(0) = q_s$ and $q(L) = q_d$. Applying the continuity conditions for q, $\partial q/\partial y$, and $\partial^2 q/\partial y^2$ at the node points in between any two charge segments and using the boundary conditions, one can derive a relation between the cubic equation coefficients a_n, b_n, c_n, and d_n for $n = 1$ to $N_{\text{SEG}} - 1$ and the charges at nodes $q_s,..., q(y = nL/N_{\text{SEG}}), ..., q_d$. For example, for $N_{\text{SEG}} = 3$, the continuity conditions to be imposed are as follows:

$$q\left(\frac{L}{3}\right)^- = q\left(\frac{L}{3}\right)^+,$$

$$q\left(\frac{2L}{3}\right)^- = q\left(\frac{2L}{3}\right)^+,$$

$$\left.\frac{\partial q}{\partial y}\right|_{y=\frac{L}{3}^-} = \left.\frac{\partial q}{\partial y}\right|_{y=\frac{L}{3}^+},$$

$$\left.\frac{\partial q}{\partial y}\right|_{y=\frac{2L}{3}^-} = \left.\frac{\partial q}{\partial y}\right|_{y=\frac{2L}{3}^+},$$

$$\left.\frac{\partial^2 q}{\partial y^2}\right|_{y=\frac{L}{3}^-} = \left.\frac{\partial^2 q}{\partial y^2}\right|_{y=\frac{L}{3}^+},$$

$$\left.\frac{\partial^2 q}{\partial y^2}\right|_{y=\frac{2L}{3}^-} = \left.\frac{\partial^2 q}{\partial y^2}\right|_{y=\frac{2L}{3}^+}. \tag{6.30}$$

Additionally, at the boundaries

$$\frac{\partial^2 q}{\partial y^2}\bigg|_{y=0} = \frac{\partial^2 q}{\partial y^2}\bigg|_{y=L} = 0. \tag{6.31}$$

The solution for the $N_{SEG} = 3$ case can be found in [10], and was derived in the context of a double gate FET charge model. The values for the so-derived coefficients remain the same. Only the function $f\left(q, \frac{\partial q}{\partial y}, \frac{\partial^2 q}{\partial y^2}\right)$ and the assumptions that go into a short-channel drain current equation change with the FET architecture.

The continuity equation at the nth node can be written as

$$\frac{\partial q_n}{\partial t} + f_n\left(q_n, \frac{\partial q_n}{\partial y}, \frac{\partial^2 q_n}{\partial y^2}\right) = 0 \tag{6.32}$$

where q_n denotes the instantaneous channel charge at the nth node. We note that the derivatives of the charge $\partial q/\partial y$ and $\partial^2 q/\partial y^2$ at the intermediate nodes can be expressed as a function of the node charges (since the coefficients a_n, etc., are a linear function of the node charges after imposing Equations (6.30) and (6.31) [11].[4] Thus the continuity equation at the nth node changes to

$$\frac{\partial q_n}{\partial t} + f_n(q_s, q_1, q_2, \ldots, q_n, \ldots, q_d) = 0. \tag{6.33}$$

$N_{SEG} - 1$ such continuity equations for all nodes can be written in a similar way. Therefore, the partial differential equation (6.28) has been converted to $N_{SEG} - 1$ ordinary differential equations. These $N_{SEG} - 1$ ordinary differential equations are coupled as the channel charge at the nth node depends on the charge at all the other nodes through the source function, $f_n()$. These $N_{SEG} - 1$ ordinary differential equations can be represented as $N_{SEG} - 1$ resistance-capacitance subcircuits within a SPICE simulator (see Figure 6.7). For SPICE implementation purposes, the resistor R_{NQS} is chosen to be a large value (1000) to facilitate convergence and the capacitance C_{NQS} is unity.

The terminal charges $Q_{d,nqs}$ and $Q_{s,nqs}$ at the drain end and the source end, respectively, can be obtained by integrating the cubic segments using the Ward-Dutton partition scheme as follows:

$$Q_{d,nqs} = -WC_{ox}\int_0^L \frac{y}{L}q(y)\,dy$$

$$Q_{d,nqs} = -WC_{ox}\left[\frac{1}{90}q_s + \frac{1}{10}q\left(\frac{L}{3}\right) + \frac{4}{15}q\left(\frac{2L}{3}\right) + \frac{11}{90}q_d\right], \tag{6.34}$$

$$Q_{s,nqs} = -WC_{ox}\int_0^L \left(1 - \frac{y}{L}\right)q(y)\,dy,$$

$$Q_{s,nqs} = -WC_{ox}\left[\frac{11}{90}q_s + \frac{4}{15}q\left(\frac{L}{3}\right) + \frac{1}{10}q\left(\frac{2L}{3}\right) + \frac{1}{90}q_d\right]. \tag{6.35}$$

---

[4]A matrix-based evaluation of the coefficients of the node charges to express the charge derivatives for any N_{SEG} can be found in [12].

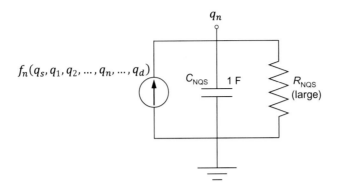

FIGURE 6.7

R-C subcircuit representation of Equation (6.33). The node voltage here represents the channel charge at the *n*th intermediate node.

The overall gate terminal charge $Q_{g,nqs}$ can then be obtained as follows:

$$Q_{g,nqs} = - \left(Q_{s,nqs} + Q_{d,nqs} + Q_{bulk} \right). \tag{6.36}$$

The above expressions for $Q_{d,nqs}$, $Q_{s,nqs}$, and $Q_{g,nqs}$ replace their QS equivalent charges derived in Section 6.1. This model can be selected by setting NQSMOD = 3.

REFERENCES

[1] Y. Tsividis, Operation and Modeling of the MOS Transistor, second ed., McGraw-Hill, New York, 1999.

[2] R.L. Pritchard, Electrical Characteristics of Transistors, McGraw-Hill, New York, 1967.

[3] S.-Y. Oh, D. Ward, R. Dutton, Transient analysis of MOS transistors, IEEE J. Solid State Circuits 15 (4) (1980) 636–643.

[4] M. Chan, K. Hui, C. Hu, P.-K. Ko, A robust and physical BSIM3 non-quasistatic transient and AC small-signal model for circuit simulation, IEEE Trans. Electron Dev. 45 (4) (1998) 834–841.

[5] Z. Zhu, G. Gildenblat, C. Mcandrew, I.-S. Lim, Accurate RTA-based nonquasistatic MOSFET model for RF and mixed-signal simulations, IEEE Trans. Electron Dev. 59 (5) (2012) 1236–1244.

[6] X. Jin, J.-J. Ou, C.-H. Chen, W. Liu, M. Deen, P. Gray, C. Hu, An effective gate resistance model for CMOS RF and noise modeling, International Electron Devices Meeting (IEDM) Technical Digest, 1998, pp. 961–964.

[7] S. Venugopalan, From Poisson to silicon—advancing compact SPICE models for IC design, Ph.D. dissertation, Dept. Elect. Eng., Univ. of California, Berkeley, CA, 2013.

[8] A. Scholten, L. Tiemeijer, P. De Vreede, D.B.M. Klaassen, A large signal non-quasi-static MOS model for RF circuit simulation, International Electron Devices Meeting (IEDM) Technical Digest, 1999, pp. 163–166.

[9] M. Bucher, A. Bazigos, An efficient channel segmentation approach for a large-signal NQS MOSFET model, Solid State Electron. 52 (2) (2008) 275–281.

[10] H. Wang, X. Li, W. Wu, G. Gildenblat, R. van Langevelde, G.D.J. Smit, A. Scholten, D.B.M. Klaassen, A unified nonquasi-static MOSFET model for large-signal and small-signal simulations, IEEE Trans. Electron Dev. 53 (9) (2006) 2035–2043.

[11] S. Sarkar, A.S. Roy, S. Mahapatra, United large and small signal non-quasi-static model for long channel symmetric DG MOSFET, Solid State Electron. 54 (11) (2010) 1421–1429.

[12] PSP 103.1 documentation [Online]. Available: http://pspmodel.asu.edu/psp_documentation.htm, 2009.

CHAPTER OUTLINE

7.1 FinFET device structure and symbol definitions 158
7.2 Modeling of geometry-dependent source/drain resistances in FinFETs 161
    7.2.1 Contact resistance 162
    7.2.2 Spreading resistance 164
    7.2.3 Extension resistance 167
7.3 Parasitic resistance model verification 169
    7.3.1 TCAD simulation setup 169
    7.3.2 Device optimization 170
    7.3.3 Extraction of source and drain resistances 172
    7.3.4 Discussion 176
7.4 Implementation considerations of the parasitic resistance model 178
    7.4.1 Physical parameters 178
    7.4.2 Resistance components 178
7.5 Gate electrode resistance model 179
7.6 FinFET parasitic capacitance models 179
    7.6.1 Connection of parasitic capacitance components 179
    7.6.2 Derivation of two-dimensional fringe capacitance 181
7.7 Modeling of FinFET fringe capacitance in three dimensions: CGEOMOD = 2 187
7.8 Parasitic capacitance model verification 188
7.9 Summary 192
References 193

In previous chapters, we discussed the modeling of intrinsic device behavior of a FinFET in BSIM-CMG. In this chapter we turn our focus to parasitic resistance and capacitance modeling. Parasitic resistance and capacitance are important components, since its magnitude is comparable to channel resistance in scaled devices. In a FinFET, parasitic resistances are difficult to model owing to the complex three-dimensional geometry.

Parasitic resistances in a FinFET include the source/drain resistance and the gate resistance. The impact of the source/drain resistance is usually more significant than the gate resistance. This is especially true with the recent introduction of the metal gate technology [1], in which highly conductive metal is used as the gate material. The 2012 International Technology Roadmap for Semiconductors [2] assumes a 33% degradation in $I_{d,sat}$ owing to source/drain series resistance for state-of-the-art MOS-FET technology. An accurate model for source/drain resistance in FinFETs is needed. The gate resistance is present owing to the finite conductivity of the gate material. At DC the gate current is very small, so the gate resistance does not alter the transistor's DC behavior. However, it does impact the AC behavior such as CMOS switching delay. A compact analytical model for gate resistances in FinFETs is available [3].

Several models for source/drain resistances in FinFETs have been proposed. Dixit et al. [4] proposed a comprehensive source/drain resistance model for a double-gate MOSFET with lithography-defined source and drain, and verified it with TCAD and measured data. Tekleab and Zeitzoff [5] extended it to consider more than one contact surface. However, these models are limited to rectangular source/drain contacts. FinFETs are expected to have several fins in parallel to have sufficient current driving capability. Research has shown that high layout density can be achieved if the multiple fins are enlarged and eventually merged using a selective epitaxial growth (SEG) process, forming a connected three-dimensional raised source/drain contact [6, 7]. The cross-section of such a raised source/drain is likely nonrectangular. This needs to be considered. Moreover, it has been shown that the performance of a fully depleted FinFET is the best with an underlaped source and drain [8, 9]. Devices with an underlaped source and drain have large bias-dependent source/drain resistance. However, both Dixit et al. [4] and Tekleab and Zeitzoff [5] considered bias-independent resistances only.

In BSIM-CMG, both bias-dependent and geometry-dependent parasitic resistance for FinFETs are modeled. We will show that the model is applicable to raised source/drain FinFETs with nonrectangular source/drain cross-section. To verify the model and demonstrate its predictivity, we will compare it with three-dimensional TCAD simulations. In addition, we will also discuss the modeling of bias-dependent source and drain resistances in the FinFET with source/drain underlap.

In this chapter, we also discuss the modeling of capacitance for BSIM-CMG. We show how the fringe capacitance is separated into gate-to-fin, gate-to-contact components in the top and sides of the fin, as well as the corners. Each component is separately derived with a semi-empirical approach and added together. We also show that the model can be generalized to single-fin as well as multi-fin cases. The model agrees well with the more accurate finite-element-based TCAD simulation. Additional parasitic capacitance elements such as junction capacitance are also modeled in BSIM-CMG with consideration of the unique FinFET geometry.

7.1 FinFET DEVICE STRUCTURE AND SYMBOL DEFINITIONS

In this study we consider FinFETs in which the selective epitaxial growth (SEG) process is applied to merge individual fins. Single-fin FinFETs and multifin FinFETs will likely coexist on the same wafer. Therefore, we assume both single-fin and

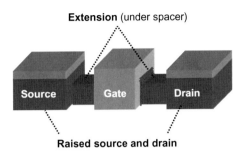

FIGURE 7.1

Bird's-eye view of a raised source/drain FinFET.

multifin FinFETs will be subject to source/drain SEG, even though fin merging is not necessary for single-fin FinFETs.

A three-dimensional drawing of the raised source/drain FinFET is shown in Figure 7.1. For simplicity we show only one fin of a multifin FinFET. The channel portion of the fin is wrapped on three sides by the gate stack. The source and drain silicon are enlarged by SEG to reduce resistances. For multifin FinFETs, a larger source and drain also makes contacting easier. The thin region not covered by the gate is the extension region. It is not subject to SEG because it is protected by the spacer (not shown in Figure 7.1) during epitaxial growth. The metallic region on top of the raised source and drain is the silicide. For single-fin FinFETs, depending on the process, the silicide may wrap around the raised source/drain on three sides.

Figure 7.2 shows a cross-sectional view of a FinFET along the source-to-drain direction. The insulating material on top of the fin and beneath the gate with height

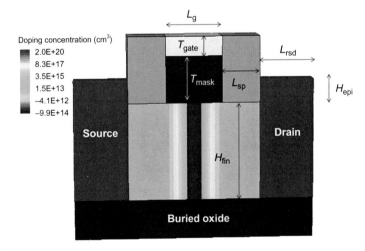

FIGURE 7.2

Cross-section of a raised source/drain double-gate FinFET and symbol definitions. This figure is generated with Sentaurus TCAD simulation tool [10].

Table 7.1 Symbol Definition

Parameter Name	Definition
L_g	Gate length
L_{sp}	Spacer thickness
L_{rsd}	Raised source/drain length
H_{fin}	Fin height
T_{gate}	Gate height
H_{epi}	Height of epitaxial silicon above fin
F_{pitch}	Fin pitch of a multifin FinFET
T_{fin}	Fin thickness
C_{ratio}	Ratio of the corner area filled with silicon to the total corner area
NFIN	Total number of fins in the FinFET
A_{rsd}	Per-fin component of the raised source/drain area
ARSDEND	End component of the raised source/drain area
DELTAPRSD	Correction term for silicide/epitaxial silicon interfacial length per fin
PRSDEND	End component of silicide/epitaxial silicon interfacial length

T_{mask} is the hard mask. In some FinFET processes, a hard mask is used for fin etching and is left on top of the fin and is never removed. Since the top surface of the fin channel does not conduct current, the device becomes a *double-gate FinFET*. In other processes, no hard mask is used. Instead $T_{mask} = T_{ox}$ and the top surface has current conduction. Such a FinFET is a *triple-gate FinFET*. FinFETs may also be classified by substrate type into silicon-on-insulator (SOI) FinFETs and bulk FinFETs. For example, Figure 7.2 shows an SOI FinFET, for which the fin is situated on top of the buried oxide. Table 7.1 lists the definition of the symbols.

SEG results in a faceted raised source/drain [6]. The final raised source/drain may look something like the drawing in Figure 7.3. The corresponding cross-sectional diagram, cut in the direction parallel to the gate, is shown in Figure 7.4.

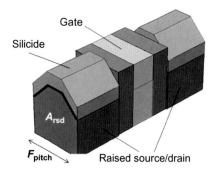

FIGURE 7.3

Bird's-eye view of a FinFET with nonrectangular source and drain epitaxy and top silicide. This figure is generated with Sentaurus TCAD simulation tool [10].

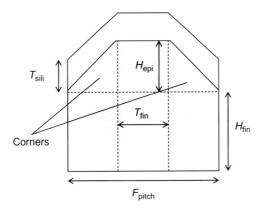

FIGURE 7.4

Two-dimensional cross-section of a FinFET with nonrectangular source and drain epitaxy and top silicide.

The cross-sectional area of the raised source/drain is A_{rsd}. For generality, the source/drain resistance is modeled as function of A_{rsd} regardless of its shape. A_{rsd} in a structure such as that shown in Figure 7.4 is given by

$$A_{rsd} = F_{pitch} \cdot H_{fin} + \left[T_{fin} + (F_{pitch} - T_{fin}) \cdot C_{ratio} \right] \cdot H_{EPI}, \quad (7.1)$$

where C_{ratio} is the ratio of the corner area filled with silicon to the total corner area. In the example given in Figure 7.4, C_{ratio} is 0.5.

Most FinFET devices in a digital circuit will have multiple fins. For multifin devices, the source/drain resistance is modeled as a function of the total area and perimeter, which are given by

$$A_{rsd,total} = A_{rsd} \times NFIN + ARSDEND, \quad (7.2)$$
$$P_{rsd,total} = (F_{pitch} + DELTAPRSD) \times NFIN + PRSDEND, \quad (7.3)$$

ARSDEND and PRSDEND are the end components associated with the first and last fins.

7.2 MODELING OF GEOMETRY-DEPENDENT SOURCE/DRAIN RESISTANCES IN FinFETs

The FinFET source/drain resistance can be separated into three components, as illustrated in Figure 7.5:

1. *Contact resistance* (R_{con}): The combined resistance due to the raised source/drain region bulk resistivity and the silicon/silicide interface resistance.
2. *Spreading resistance* (R_{sp}): The resistance due to current spreading from the source/drain extension into the raised source/drain.
3. *Extension resistance* (R_{ext}): The bias-dependent resistance in the thin source/drain extension region under the spacer.

We will discuss about each of these components in the following three subsections.

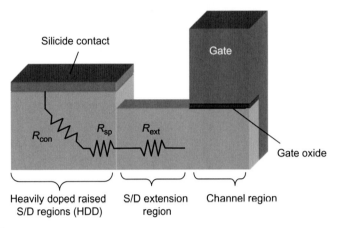

FIGURE 7.5

Separation of FinFET source/drain resistance into three components: R_{con}, contact resistance; R_{sp}, spreading resistance; and R_{ext}, source/drain extension resistance.

7.2.1 CONTACT RESISTANCE

The contact resistance model accounts for both the bulk resistivity in the raised source/drain region and contact resistance at the silicon/silicide interface. Since the resistance is distributed, it is difficult to separate the two into individual resistors.

To consider the distributed effect, we partition the raised source/drain region into infinitesimally thin vertical slices (Figure 7.6a). The slices are connected in a resistance network as shown in Figure 7.6b. For each slice, there is a bulk resistance component, ΔR_s, between adjacent slices, and a contact resistance component, ΔR_c, from each slice to the contact. The bulk resistance component is given by

$$\Delta R_s = \rho \cdot \frac{\Delta x}{H_{rsd} \cdot W_{rsd}}, \tag{7.4}$$

where ρ is the bulk resistivity, which is given by

$$\rho = \frac{1}{q \cdot N_{rsd} \cdot \mu_{rsd}}. \tag{7.5}$$

N_{rsd} is the raised source/drain region doping concentration. We assume the raised source/drain region is *in situ* doped during SEG and is uniformly doped. The mobility μ_{rsd} is calculated using Masetti's model [11] as function of N_{rsd}. The contact resistance component is given by

$$\Delta R_c = \frac{\rho_c}{\Delta x \cdot W_{rsd}}, \tag{7.6}$$

where ρ_c is the specific contact resistivity in units of Ω-cm^2.

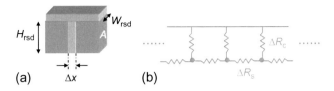

(a) Δx (b)

FIGURE 7.6

(a) An infinitesimal slice of the source/drain considered for distributed contact resistance derivation. (b) Equivalent resistance network for distributed contact resistance calculation.

Equations (7.4) and (7.6) are valid only for rectangular contacts. To generalize them to any contact shape and multifin devices, we express them in terms of the raised source/drain cross-sectional area and the interface peripheral length:

$$\begin{cases} \Delta R_s = \rho \cdot \dfrac{\Delta x}{A_{rsd,total}}, \\[3mm] \Delta R_c = \dfrac{\rho_c}{\Delta x \cdot P_{rsd,total}}. \end{cases} \qquad (7.7)$$

The transmission line model [12] is applied to solve this problem. By solving a differential equation, we obtain the total contact resistance:

$$R_{con} = \rho \cdot \frac{L_T}{A_{rsd,total}} \cdot \frac{\eta \cdot \cosh \alpha + \sinh \alpha}{\eta \cdot \sinh \alpha + \cosh \alpha}, \qquad (7.8)$$

where

$$L_T = \sqrt{\frac{\rho_c \cdot A_{rsd,total}}{\rho \cdot P_{rsd,total}}}, \qquad (7.9)$$

$$\alpha = \frac{L_{rsd}}{L_T}, \qquad (7.10)$$

$$\eta = \frac{\rho_c \cdot A_{rsd,total}}{\rho \cdot L_T \cdot A_{term}}. \qquad (7.11)$$

A_{term} is the silicon/silicide area at the two ends of the FinFET for a structure such as that shown in Figure 7.7. A special case is for a FinFET without the end contacts, so that $A_{term} = 0$. Equation (7.8) reduces to

$$R_{con} = \rho \cdot \frac{L_T}{A_{rsd,total}} \cdot \coth \alpha \qquad (7.12)$$

Note that Equation (7.12) is similar to but more general than the contact resistance formula in [4, 5], since it can model a nonrectangular raised source/drain cross-sectional geometry.

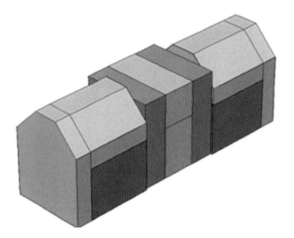

FIGURE 7.7

FinFET with a nonrectangular epitaxial source/drain and silicide on the top and at the two ends. This figure is generated with Sentaurus TCAD simulation tool [10].

7.2.2 SPREADING RESISTANCE

When current flows from the source/drain extension region into the raised source/drain region, it spreads out gradually. We model the resistance increase due to such spreading as a new component, the spreading resistance. This spreading phenomenon is also known as current crowding.

The top view of the raised source/drain and extension regions is shown in Figure 7.8, with the gray area representing the region through which the current flows. We assume that the current spreading boundary is at an angle θ to the fin direction.

We first consider a case where the cross-sections of the extension and the raised source/drain are both squares. In between, each current flow cross-section is also a square. We assume the side length of the squares increases linearly with position. Consequently each slice in the spreading region with thickness Δx has resistance of

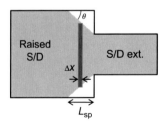

FIGURE 7.8

Illustration of current spreading path from the source/drain extension region into the raised source/drain region.

$$\Delta R = \frac{\rho \cdot \Delta x}{\left(\sqrt{A_{\text{fin}}} + 2 \cdot x \cdot \tan\theta\right)^2},$$ (7.13)

where A_{fin} is the cross-sectional area of the fin extension, which is

$$A_{\text{fin}} = H_{\text{fin}} \cdot T_{\text{fin}}.$$ (7.14)

The resistance in the spreading resistance is given by an integral from 0 to L_1:

$$R = \int_0^{L_1} \frac{\rho \cdot dx}{\left(\sqrt{A_{\text{fin}}} + 2 \cdot x \cdot \tan\theta\right)^2},$$ (7.15)

where L_1 satisfies the relation

$$2 \cdot L_1 \cdot \tan\theta = \sqrt{A_{\text{rsd}}} - \sqrt{A_{\text{fin}}}.$$ (7.16)

After carrying out the integration, we obtain the total resistance in the spreading region:

$$R = \frac{\rho \cdot \cot\theta}{2} \left(\frac{1}{\sqrt{A_{\text{fin}}}} - \frac{1}{\sqrt{A_{\text{rsd}}}} \right).$$ (7.17)

If we carry out the same analysis for a circular fin extension and a circular raised source/drain, we will obtain a similar result:

$$R = \frac{\rho \cdot \cot\theta}{\sqrt{\pi}} \left(\frac{1}{\sqrt{A_{\text{fin}}}} - \frac{1}{\sqrt{A_{\text{rsd}}}} \right).$$ (7.18)

To be more general we express the resistance in the spreading region as

$$R = \frac{\rho \cdot \cot\theta}{s} \left(\frac{1}{\sqrt{A_{\text{fin}}}} - \frac{1}{\sqrt{A_{\text{rsd}}}} \right),$$ (7.19)

where the shape parameter s depends on the shape of the fin extension and the raised source/drain.

We can also calculate R', the total resistance in the same region if there were no spreading:

$$R' = \frac{\rho \cdot L_1}{A_{\text{rsd}}} = \frac{\rho \cdot \cot\theta}{s \cdot A_{\text{rsd}}} \left(\sqrt{A_{\text{rsd}}} - \sqrt{A_{\text{fin}}} \right).$$ (7.20)

Since the spreading resistance is defined as the increase in resistance, it is the difference between R and R', which is

$$R_{\text{sp}} = \frac{\rho \cdot \cot\theta}{s} \cdot \left(\frac{1}{\sqrt{A_{\text{fin}}}} - \frac{2}{\sqrt{A_{\text{rsd}}}} + \frac{\sqrt{A_{\text{fin}}}}{A_{\text{rsd}}} \right).$$ (7.21)

We define

$$R_0 = \rho \left(\frac{1}{\sqrt{A_{\text{fin}}}} - \frac{2}{\sqrt{A_{\text{rsd}}}} + \frac{\sqrt{A_{\text{fin}}}}{A_{\text{rsd}}} \right)$$ (7.22)

and let

$$R_{\text{sp}} = K \cdot R_0,$$ (7.23)

where the slope factor K is

$$K = \frac{\cot\theta}{s}.$$ (7.24)

We hypothesize that K is insensitive to the device geometry in the range we are interested in.

To test the hypothesis, three-dimensional TCAD simulations are performed to compute R_{sp}. We simulate test structures that consist of a uniformly doped silicon block with contacts on both sides, as illustrated in Figure 7.9a. The specific contact resistivity is set to a very small value so that its effect is negligible.[1] The doping concentration is set to $N_{sd} = 2 \times 10^{20}$ cm^{-3}. The dimensions of the left-side contact are fixed at $H_{rsd} = 60$ nm and $W_{rsd} = 45$ nm. The height of the right-side contact varies from $H_{fin} = 30$ nm to $H_{fin} = 60$ nm in steps of 5 nm, and the width varies simultaneously from $T_{fin} = 15$ nm to $T_{fin} = 45$ nm in steps of 5 nm. For each case five different raised source/drain length, L_{rsd}, are simulated. The spreading resistance

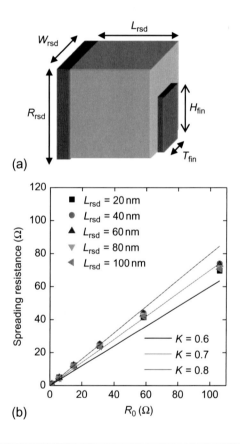

(a)

(b)

FIGURE 7.9

Extraction of spreading resistance: (a) the test structure and (b) extraction of the slope factor, K.

[1] ρ_c is set to 10^{-12} Ω-cm^2 in the TCAD simulation. In TCAD tools, we are generally not allowed to set ρ_c to zero.

is extracted by subtracting out the resistance of the nonspreading case, where the nonspreading case is when $W_{rsd} = T_{fin}$ and $H_{rsd} = H_{fin}$.

In Figure 7.9b we have plotted the spreading resistance versus R_0 for both Equation (7.23) and TCAD simulation results. The best agreement is obtained at $K = 0.7$. The model and TCAD data agree reasonably well, suggesting that the constant K is a reasonable assumption.

7.2.3 EXTENSION RESISTANCE

Extension resistance (R_{ext}) refers to the resistance in the fin extension region under the spacer. The modeling of extension resistance requires knowledge of the doping profile in the extension region, which is often not accurately known in reality. The profile shape varies depending on the process condition. In addition, surface accumulation due to the fringe field originating from the gate creates bias dependency.

To simplify the problem, we make several assumptions about the spacer configuration and doping profile, as illustrated in Figure 7.10. We assume the spacer (with length L_{sp}) consists of an offset spacer (with length L_{off}) and a main spacer. Implantation is performed after the offset spacer is deposited but before the main spacer forms. As a result, the doping is uniform under the main spacer, but decays following a Gaussian profile under the offset spacer. N_{ext} is the doping concentration at the offset spacer edge. Although an implanted extension is assumed, this model can also be applied to FinFETs with extension doping coming solely from an in situ doped epitaxial source/drain.

The extension resistance is modeled as one bias-dependent accumulation resistance component and two bias-independent bulk resistance components, and these are combined into a resistance network as shown in Figure 7.11.

The accumulation resistance (R_{acc}) represents the conductive path at the surface of the source/drain extension due to charge accumulation induced by the gate fringe

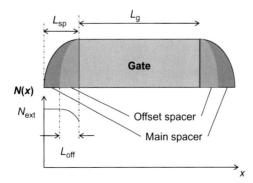

FIGURE 7.10

Doping profile and spacer configurations considered for R_{ext} modeling.

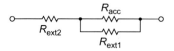

FIGURE 7.11

Subcircuit for resistance modeling in the accumulation region.

fields [4]. The accumulation resistance is significant and needs to be properly considered, especially for devices with little or no source/drain to gate overlap and that have a relatively small doping concentration at the gate edge. The accumulation resistance is modeled using the following expression:

$$R_{\mathrm{acc}} = \frac{R_{\mathrm{acc0}}}{H_{\mathrm{fin}} \cdot (V_{\mathrm{gs(d)}} - V_{\mathrm{fbsd}})}, \tag{7.25}$$

where V_{fbsd} is the flatband voltage at the source and drain. The accumulation resistance is inversely proportional to the conduction charge density, which is proportional to $V_{\mathrm{gs(d)}} - V_{\mathrm{fbsd}}$.

The bulk resistance of the fin extension is modeled as two separate components: R_{ext1} and R_{ext2}. R_{ext1} represents the bulk resistance of the fin beneath the accumulated part of the source/drain. We assume it is partially under the offset spacer and partially under the main spacer. R_{ext1} is given by

$$R_{\mathrm{ext1}} = \frac{R_{\mathrm{ext1,0}}}{H_{\mathrm{fin}} T_{\mathrm{fin}}}. \tag{7.26}$$

We have lumped the complex doping profile and mobility change due to doping into the variable $R_{\mathrm{ext1,0}}$.

Further away from the gate, there is no accumulation at the surface but only conductivity in the bulk of the fin. We model it with the following expression:

$$R_{\mathrm{ext2}} = \frac{R_{\mathrm{ext2,0}} \cdot (L_{\mathrm{sp}} - \Delta L_{\mathrm{ext}})}{H_{\mathrm{fin}} T_{\mathrm{fin}}}, \tag{7.27}$$

where we have assumed R_{ext2} is located under the main spacer where the doping is horizontally uniform.

After combining the three resistance components into a network as shown in Figure 7.11, we obtain the full expression for extension resistance:

$$R_{\mathrm{ext}} = \left(\frac{R_{\mathrm{acc0}}}{H_{\mathrm{fin}}(V_{\mathrm{gs(d)}} - V_{\mathrm{fbsd}})} \right) \left\| \left(\frac{R_{\mathrm{ext1,0}}}{H_{\mathrm{fin}} T_{\mathrm{fin}}} \right) + \frac{R_{\mathrm{ext2,0}} \cdot (L_{\mathrm{sp}} - \Delta L_{\mathrm{ext}})}{H_{\mathrm{fin}} T_{\mathrm{fin}}} \right. . \tag{7.28}$$

The above equation can be simplified to

$$R_{\mathrm{ext}} = \frac{\frac{R_{\mathrm{ext1,0}}}{H_{\mathrm{fin}} T_{\mathrm{fin}}}}{1 + \frac{R_{\mathrm{ext1,0}}}{T_{\mathrm{fin}} R_{\mathrm{acc0}}} \cdot (V_{\mathrm{gs(d)}} - V_{\mathrm{fbsd}})} + \frac{R_{\mathrm{ext2,0}} \cdot (L_{\mathrm{sp}} - \Delta L_{\mathrm{ext}})}{H_{\mathrm{fin}} T_{\mathrm{fin}}}, \tag{7.29}$$

which, interestingly, turns out to have the same bias dependence as the BSIM4 model [13][2]:

$$R_s = \frac{1}{W_{eff}} \left(\frac{RSW}{1 + PRWG \cdot (V_{gs} - V_{fbsd})} + RSWMIN \right), \tag{7.30a}$$

$$R_d = \frac{1}{W_{eff}} \left(\frac{RDW}{1 + PRWG \cdot (V_{gd} - V_{fbsd})} + RDWMIN \right). \tag{7.30b}$$

7.3 PARASITIC RESISTANCE MODEL VERIFICATION

In this section, verification of the model with three-dimensional TCAD simulation is presented. The details of the TCAD simulation setup, the separation of bias-dependent and bias-independent source/drain resistances, and the comparison of the model versus TCAD simulation are discussed. In addition, we present our findings that the traditional transmission line method overestimates L_{eff}, and the physical L_{eff} must be extracted by other means.

7.3.1 TCAD SIMULATION SETUP

Three-dimensional numerical simulations of the FinFET are carried out using TCAD. In this section the details of the simulation setup are presented.

The simulation grid is created with Sentaurus Structure Editor [14] and the mesh generation program Noffset3D [15]. A bird's-eye view of the FinFET simulation structure is shown in Figure 7.12. Because of symmetry, only one-half of the FinFET needs to be simulated. This reduces the total number of grid points by half and speeds up the simulation with no change in accuracy. The spacer is intentionally made transparent to show the fin extension region. The raised source/drain has a wrapped-around contact. The top and sides of the raised source/drain as well as the source/drain end planes are in contact with silicide. Such contact is possible for FinFETs with an unmerged epitaxial source/drain or SRAM FinFETs with a single fin per device. With a hard mask on top of the fin, the structure is a double-gate FinFET. Moreover, the device is situated on top of the buried oxide; therefore, it is an SOI FinFET. The nominal geometry considered in this study is a linearly scaled version of that of a manufacturable FinFET technology [16]. Table 7.2 lists the nominal FinFET geometry and other simulation parameters for TCAD simulation.

The current-voltage characteristics are simulated using Sentaurus Device [10]. Doping-dependent mobility is modeled using the Masseti mode [11]. Mobility degradation at high vertical fields is accounted for with the enhanced Lombardi model [10, 17]. Velocity saturation is modeled using the extended Canali model

---

[2]PRWB is set to 0. We consider the case when RDSMOD = 1, which activates the external resistance model.

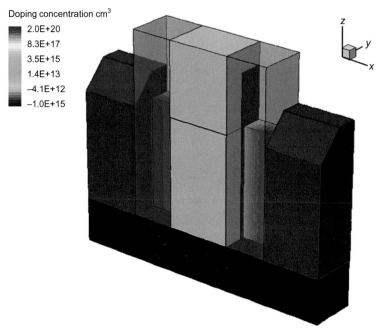

Doping concentration cm^3

■	2.0E+20
	8.3E+17
	3.5E+15
	1.4E+13
	−4.1E+12
■	−1.0E+15

FIGURE 7.12

Bird's eye view of the FinFET structure for TCAD simulation. The nitride spacer is intentionally made semitransparent to make the fin extension visible.

[10, 18]. Quantum mechanical effects are taken into account with an equivalent oxide thickness (EOT) approach. In other words, we lump the inversion layer thickness as part of the EOT. The impact of quantum effects on source/drain Schottky barrier height lowering [19] is lumped into the specific contact resistivity (ρ_c). For the nominal case we set $\rho_c = 10^{-8}$ Ω-cm^2, as reported by the International Technology Roadmap for Semiconductors [20]. Lower contact resistivity values may be achievable [19, 21]. Since the analytical model we developed is scalable, it can model those low ρ_c values as well.

7.3.2 DEVICE OPTIMIZATION

We use a metal gate in the simulation, and assume that threshold voltage (V_{th}) tuning through gate work function engineering is possible. We optimize the FinFET doping profile for maximum drive current (I_{on}) at a constant I_{off}/W_{eff} of 1 nA/μm by altering the extension doping, N_{ext}, and the offset spacer width, L_{off} (Figure 7.13). For all extension doping concentrations, I_{on} versus L_{off} is a bell-shaped curve. When the offset spacer is too thin, the extension doping encroaches into the channel and degrades the subthreshold swing, forcing V_{th} to be very high for the same I_{off} and reducing I_{on}. On the other hand, when the offset spacer is too thick, the

Table 7.2 Nominal FinFET Geometry and Other Simulation Parameters Used for TCAD Simulation

Parameter Name	Description	Nominal Value
L_g	Physical gate length	15 nm
EOT	Equivalent oxide thickness	1.0 nm
T_{fin}	Fin thickness	10 nm
H_{fin}	Fin height	25 nm
H_{rsd}	Raised source/drain height	31 nm
T_{mask}	Oxide hard mask thickness	12 nm
L_{off}	Offset spacer width	6 nm
L_{rsd}	Raised source/drain length	14 nm
L_{sp}	Source/drain spacer width	10 nm
T_{epi}	Horizontal epi thickness	6 nm
N_{ext}	Source/drain extension doping	2×10^{20} cm^{-3}
N_{rsd}	Raised source/drain epi doping	2×10^{20} cm^{-3}
N_{body}	Fin body doping	10^{15} cm^{-3}
LDG	Lateral doping gradient in extension	2.5 nm/dec
ρ_c	Contact resistivity	10^{-8} Ω-cm^2
V_{dd}	Supply voltage	0.9 V

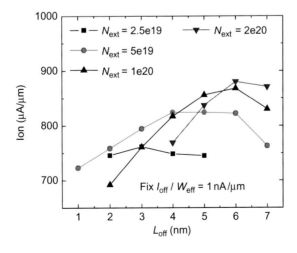

FIGURE 7.13

Optimization of the offset spacer width (L_{off}) and fin extension doping (N_{ext}) for maximum on current.

effective channel length becomes very large, the source/drain resistance becomes significant, and on current is degraded as well. Maximum I_{on} is achieved with $N_{ext} = 2 \times 10^{20}\,\text{cm}^{-3}$ and $L_{off} = 6\,\text{nm}$ over the range of doping and offset spacer simulated. The corresponding metal gate work function for $1\,\text{nA}/\mu\text{m}$ off current is 4.603 eV.

As we will show later, with $L_{off} = 6\,\text{nm}$, L_{eff} is a function of gate overdrive, and ranges from about 16 to 20 nm for $V_{gs} < V_{dd}$. Therefore, the optimal device has an underlapped source/drain design. This conclusion is consistent with [22, 23], where the optimal design of a FinFET source/drain is underlapped. We are assuming here that band-to-band tunneling leakage (or gate-induced drain leakage) is small enough and its effect on I_{off} can be neglected.

7.3.3 EXTRACTION OF SOURCE AND DRAIN RESISTANCES

L_{eff} and the source/drain series resistance are often extracted by the transmission line method by plotting the channel resistance $R_{total} = V_{ds}/I_d$ versus the design gate length L_{des} at several gate overdrive values ($V_{gs} - V_{th}$) at $V_{ds} = 50\,\text{mV}$ and finding their interception point [24]. There are two potential issues:

1. The average channel mobility is a function of the gate length owing to, for example, an angled halo implant near the source/drain that suppresses subsurface leakage current.
2. The source/drain resistance is a function of gate bias.

The former is less of an issue for fully depleted FinFETs since a halo implant is usually unnecessary for FinFETs. On the other hand, the source/drain resistance has significant bias dependence since the conductivity of the fin extension near the gate edge is modulated by the gate fringe field. The bias dependence is expected to be more significant for devices with little or no source/drain overlap, as is the nominal case in this study. As a consequence, L_{eff} and R_{ds} extracted using the transmission line method are likely to be quite different from their physical values. To illustrate this, we performed a TCAD simulation for R_{total} versus L_g and fitted the results to linear curves (Figure 7.14). The extrapolated curves intersect at approximately $L_g = -50\,\text{nm}$. If we had assumed there is no bias dependence, we would have concluded there is a 25 nm underlap on each side, which is unlikely given the device structure we have. Moreover, the intersection point itself is a function of bias.

In this study an alternative TCAD-based method for source/drain resistance extraction is used. The total channel resistance is given by

$$R_{total}(V_{gs}) = R_{ds}(V_{gs}) + \frac{L_g - \Delta L(V_{gs})}{\mu C_{ox} W_{eff}(V_{gs} - V_{th} - \frac{V_{ds}}{2})V_{ds}}. \tag{7.31}$$

While it is difficult to extract $R_{ds}(V_{gs})$ and $\Delta L(V_{gs})$ simultaneously, if $\Delta L(V_{gs})$ is known we can find $R_{ds}(V_{gs})$ as the zero crossing of the R_{total} versus L_g curve. $\Delta L(V_{gs})$, defined as the point at which the electron concentration at the inversion

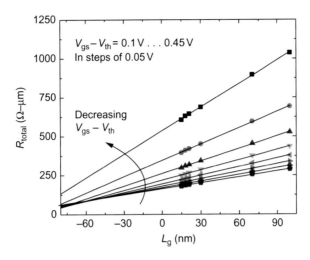

FIGURE 7.14

The total channel resistance R_{total} versus physical gate length L_g at several values of gate overdrive, $V_{gs} - V_{th}$, for the optimized device described in Section 7.3.2. V_{th} is obtained by extrapolating $I_d - V_{gs}$ at the point with maximum g_m.

charge centroid is equal to the background doping, can be extracted from the TCAD simulation. This extraction method is carried out, and the results are listed in Table 7.3.

L_{eff} is plotted versus gate overdrive, $V_{gs} - V_{th}$, in Figure 7.15 for a 15 nm device. To smoothen out numerical error, we conduct a second-order polynomial fit, which gives

Table 7.3 Extraction of overlap length (ΔL) from TCAD

V_{gs} (V)	$V_{gs} - V_{th}$ (V)	X_{DC} (nm)	$n_e(X_{DC})$ (cm^{-3})	L_{eff} (nm)	ΔL (nm)
0.45	0.042	2.43	5.85×10^{17}	15.88	−0.88
0.495	0.087	2.22	1.09×10^{18}	16.51	−1.51
0.54	0.132	1.99	1.81×10^{18}	16.98	−1.98
0.585	0.177	1.76	2.85×10^{18}	17.58	−2.58
0.63	0.222	1.55	4.21×10^{18}	18.02	−3.02
0.675	0.267	1.37	5.87×10^{18}	18.49	−3.49
0.72	0.312	1.22	7.76×10^{18}	18.84	−3.84
0.765	0.357	1.09	9.83×10^{18}	19.11	−4.11
0.81	0.402	0.99	1.21×10^{19}	19.35	−4.35
0.855	0.447	0.91	1.55×10^{19}	19.69	−4.69
0.9	0.492	0.83	1.93×10^{19}	20.08	−5.08

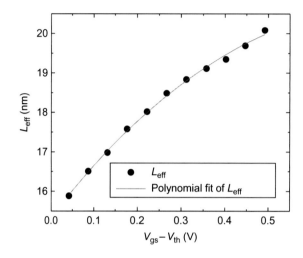

FIGURE 7.15

Effective channel length as a function of gate overdrive.

$$L_{eff} \text{ (nm)} = 15.34 + 14.06 \cdot V_{gt} - 9.42 \cdot V_{gt}^2, \tag{7.32}$$

or,

$$\Delta L \text{ (nm)} = 0.34 + 14.06 \cdot V_{gt} - 9.42 \cdot V_{gt}^2. \tag{7.33}$$

With this model of $\Delta L(V_g)$, we can now calculate $R_{ds}(V_g)$. Figure 7.16 shows R_{ds} versus V_g for two cases. For one case, we apply Equation (7.33) and find the

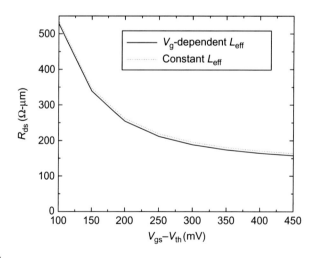

FIGURE 7.16

Source/drain resistance as a function of gate overdrive.

total source and drain resistance at $L_g = \Delta L$ (V_g-dependent L_{eff}). For the other case, we assume $\Delta L = 0$ (constant L_{eff}). The results are similar, suggesting that the bias dependence of ΔL is not a significant factor for determining R_{ds}. From now on, we will extract the source/drain resistance using $\Delta L = 0$, which is perhaps more realistic since ΔL is not known exactly in experiments.

In Section 7.2.3 we showed that the bias-dependent source/drain resistance is given as Equation (7.30a). We further assume RSW = RDW and RSWMIN = RDWMIN for symmetry reasons and carry out fitting, which gives

$$R_{ds}\ (\Omega\text{-}\mu\text{m}) = 107.5 + \frac{95.0}{1 + 7.54 \cdot (V_{gs} - V_{fbsd})} + \frac{95.0}{1 + 7.54 \cdot (V_{gd} - V_{fbsd})}. \tag{7.34}$$

The above expression fits the TCAD data very well from $V_{gs} - V_{th} = 0.1$ V to $V_{gs} - V_{th} = 4.2$ V, suggesting the resistance network in Figure 7.11 is a good description of the resistances in the extension region. The model fitting and TCAD results are plotted together in Figure 7.17, focusing on the low-V_{gs} part.

The parameters $R_{ext1,0}$ and $R_{acc,0}$ can be extracted from the bias-dependent part of Equation (7.34). On the other hand, the bias-independent part (with a value of 107.5) includes the extension resistance (R_{ext2}), the spreading resistance, and the contact resistance. The parameters $R_{ext2,0}$ and ΔL_{ext} can be extracted by the linear fitting of R_{ds} versus the spacer thickness (L_{sp}), as shown in Figure 7.18. The model fits the TCAD results very well. The extracted model parameters are summarized in Table 7.4.

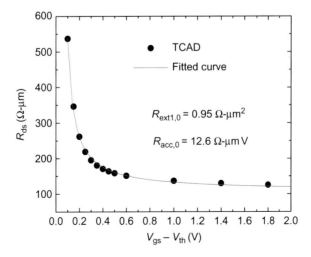

FIGURE 7.17

Source/drain resistance as a function of gate overdrive ($V_{th} = 0.408$ V).

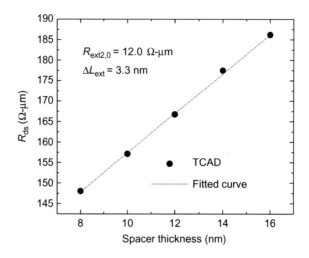

FIGURE 7.18

Source/drain resistance as a function of spacer thickness at $V_{gs} = 0.9\,V$, $V_{ds} = 50\,mV$. The channel resistance has been subtracted out.

Table 7.4 Summary of Extracted Parameters

Parameter Name	Value	Units
$R_{ext1,0}$	0.95	Ω-μm^2
$R_{acc,0}$	12.6	Ω-$\mu m\,V$
$R_{ext2,0}$	12.0	Ω-μm
ΔL_{ext}	3.3	nm

To verify Equation (7.8), we also plot the total source/drain resistance versus the raised source/drain length (L_{rsd}) in Figure 7.19. The model agrees with the TCAD simulation well without the need to introduce a fitting parameter for the contact resistance.

7.3.4 DISCUSSION

On the basis of the fitted model, individual resistance components are plotted (Figure 7.20). For the nominal case (with specific contact resistivity $\rho_c = 10^{-8}\,\Omega$-cm^2), the extension resistance is larger than the contact resistance even at $V_{gs} = V_{dd}$. This is due to the underlapped source/drain junction design with large extension resistance considered in this study.

Figure 7.20 also suggests that the spreading resistance is a relatively small component. Therefore, the error introduced with constant angle approximation should not have a significant impact on the overall result.

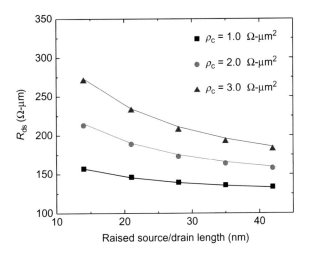

FIGURE 7.19

Source/drain resistance as a function of raised source/drain length at $V_{gs} = 0.9\,V$, $V_{ds} = 50$ mV. The channel resistance has been subtracted out.

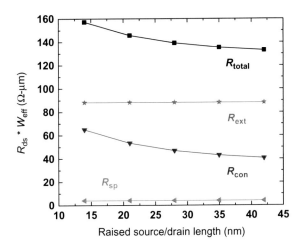

FIGURE 7.20

Breakdown of source/drain resistance into individual components: contact resistance (R_{con}), extension resistance (R_{ext}), and spreading resistance (R_{sp}).

The saturation of contact resistance at around 30 nm raised source/drain length (Figures 7.19 and 7.20) suggests that further lengthening the source/drain has less benefit. In practice, the gate pitch is fixed for a given technology, and the room for raised source/drain length optimization is limited.

7.4 IMPLEMENTATION CONSIDERATIONS OF THE PARASITIC RESISTANCE MODEL

In this section we describe the practical implementation considerations of the contact, spreading, and extension resistance models in BSIM-CMG.

7.4.1 PHYSICAL PARAMETERS

The bulk resistivity parameter ρ affects both the contact and the spreading resistance models. The user can specify ρ via the parameter RHORSD. Otherwise, BSIM-CMG computes ρ using Masetti's model [11] as described in Section 7.2.1.

The formulations of the fin cross-sectional area A_{fin} and the raised source and drain cross-sectional area A_{rsd} are implemented in BSIM-CMG as described in Section 7.1. Special consideration is given to the rare case where the top of the epitaxial source/drain is below the top of the fin ($H_{\text{epi}} < 0$). In such a case A_{fin} is reduced to $T_{\text{fin}} \times (H_{\text{fin}} + H_{\text{epi}})$.

7.4.2 RESISTANCE COMPONENTS

Since the FinFET extension resistance model has the same mathematical form as the planar BSIM4 model, we adopt the same parameter naming conventions (rather than R_{ext1}, R_{ext2}, R_{ecc}, etc., in Section 7.2.3).

Like BSIM4, two options for source/drain resistance implementation are provided in BSIM-CMG. If RDSMOD = 0, the extension resistance is implemented as a current degradation factor that is multiplied by the drain current.[3] On the other hand, if RDSMOD = 1, the extension resistance is implemented as two resistors in series with the transistor.

To be consistent with prior BSIM models, we provide users with the option to set RGEOMOD = 0 to specify geometry-dependent parasitic source/drain resistances via the sheet resistance parameters RSHS and RSHD, where

$$R_{\text{s,geo}} = \text{NRS} \cdot \text{RSHS}, \tag{7.35}$$

$$R_{\text{d,geo}} = \text{NRD} \cdot \text{RSHD}. \tag{7.36}$$

In such a case the spreading and contact resistance models described in this chapter are not used for calculations.

If, on the other hand, the user sets RGEOMOD = 1, the spreading and contact resistance formulations described in this chapter will be used.

The spreading resistance model in BSIM-CMG is implemented as Equation (7.23). The default K is set to about 0.4, which corresponds to $\theta = 55°$. This value originated from an early study and is kept unchanged for backward compatibility reasons.

---

[3]For RDSMOD = 0, the gate bias dependence term is modified from $V_{\text{gs(d)}} - V_{\text{fbsd}}$ to the average channel inversion charge density q_{ia}. This is to avoid a hump in drain current behavior, which may appear because $V_{\text{gs(d)}} - V_{\text{fbsd}}$ with RDSMOD = 0 is forced to zero via a smoothing function when it is negative.

The contact resistance model in BSIM-CMG is implemented as Equation (7.8). The raised source/drain area and the end contact are assumed to be the same ($A_{\text{rsd}} = A_{\text{term}}$) if the user sets SDTERM $= 1$. Otherwise, if SDTERM $= 0$, no end contact is considered and the contact resistance formulation falls to Equation (7.12).

Whether RGEOMOD $= 0$ or RGEOMOD $= 1$, $R_{\text{s,geo}}$ and $R_{\text{d,geo}}$ are implemented as resistors in series with the transistor. Note that if RGEOMOD $= 1$, five fitting parameters are introduced to capture extra geometry dependency [25].

7.5 GATE ELECTRODE RESISTANCE MODEL

The gate electrode resistance model can be switched on by setting RGATEMOD $= 1$. This introduces an internal node "ge." The gate electrode resistor (R_{geltd}) is placed between the external "g" node and the internal "ge" node.

The gate electrode resistance model takes into account the number of gate contacts, NGCON. NGCON $= 1$ indicates single-sided contact; NGCON $= 2$ indicates double-sided contact. R_{geltd} is given by

$$R_{\text{geltd}} = \begin{cases} \frac{\text{RGEXT}+\text{RGFIN}\cdot\text{NFIN}/3}{\text{NF}} & \text{for NGCON} = 1, \\ \frac{\text{RGEXT}/2+\text{RGFIN}\cdot\text{NFIN}/12}{\text{NF}} & \text{for NGCON} = 2. \end{cases} \tag{7.37}$$

7.6 FinFET PARASITIC CAPACITANCE MODELS

In addition to the intrinsic capacitance as described in Chapter 8, BSIM-CMG models both the overlap capacitance (C_{ov}), which is bias dependent, and the fringe capacitance (C_{fr}), which typically has weak or negligible bias dependence. We often use C_{ov} to capture the bias-dependent C-V characteristics of a given device over the bias range of interest. On the other hand, C_{fr} provides scaling of the capacitance across device geometries. For C_{fr}, BSIM-CMG offers three models with different levels of complexity, which users can select via the model parameter CGEOMOD. In this chapter, we focus on CGEOMOD $= 2$, which is the most complete and physics-based model. Its superior scalability with respect to geometrical parameters as compared with CGEOMOD $= 0$ and CGEOMOD $= 1$ makes it better suited for device variability modeling.

7.6.1 CONNECTION OF PARASITIC CAPACITANCE COMPONENTS

Both C_{fr} and C_{ov} have gate-to-source and gate-to-drain components. For CGEOMOD $= 0$ and CGEOMOD $= 2$, the respective fringe and overlap components are lumped together and connected between the gate and the inner source or drain nodes, as illustrated in Figure 7.21.

To offer modeling flexibility, a special case is CGEOMOD $= 1$, where the fringe components are connected to the outer source/drain nodes, whereas the overlap components, being closer to the channel, are connected to the inner source/drain nodes (Figure 7.22).

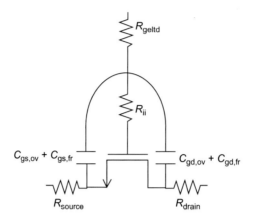

FIGURE 7.21

R-C network for CGEOMOD $= 0$ and CGEOMOD $= 2$. In this illustration the non-quasi-static resistance (R_{ii}), gate electrode resistance (R_{geltd}), and external source and drain resistances (R_{source} and R_{drain}) are all turned on by setting NQSMOD $= 1$, RGATEMOD $= 1$, and RDSMOD $= 1$.

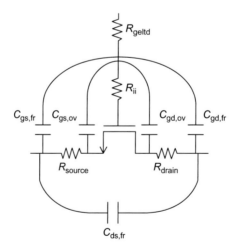

FIGURE 7.22

R-C network for CGEOMOD $= 1$. In this illustration the non-quasi-static resistance (R_{ii}), gate electrode resistance (R_{geltd}), and external source and drain resistances (R_{source} and R_{drain}) are all turned on by setting NQSMOD $= 1$, RGATEMOD $= 1$, and RDSMOD $= 1$.

Details of CGEOMOD = 0 and CGEOMOD = 1 can be found in the BSIM-CMG technical manual [25].

7.6.2 DERIVATION OF TWO-DIMENSIONAL FRINGE CAPACITANCE

In planar MOSFETs, perpendicular surfaces of the gate edge and the source/drain allow a simple conformal mapping equation to calculate the fringe capacitance (e.g., [26]). In a FinFET, however, to the best of our knowledge, no known exact two-dimensional conformal mapping method exists, given a raised source and drain structure as illustrated in Figure 7.23.

Considering Figure 7.23, we adopt a solution where the total capacitance is decomposed into a fin-to-gate component, C_{fg}, and a fin-to-channel component, C_{cg}. C_{cg} is further separated into C_{cg1}, C_{cg2}, and C_{cg3}, which have distinct electric field line trajectories (Figure 7.24). Other parasitic components are expected to be relatively small.

Each component is calculated by summing infinitesimal capacitors, each having a capacitance of

$$\Delta C = \epsilon_{ox} \frac{\Delta A}{d},\tag{7.38}$$

where ΔA is the area of the infinitesimal capacitor, and d is the length of the electric field line.

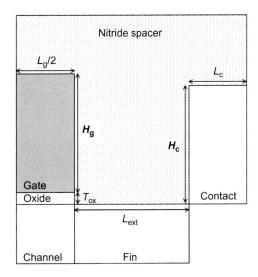

FIGURE 7.23

Two-dimensional half-device cross-section for the FinFET, where the potential distribution in the nitride region must be solved to derive the capacitance.

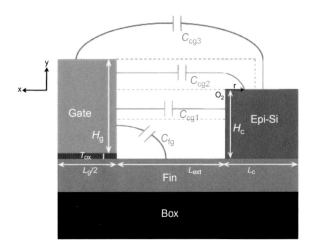

FIGURE 7.24

Separation of fringe capacitance components into fin-to-gate (C_{fg}) and contact-to-gate (C_{gd1}, C_{cg2}, and C_{cg3}) components.

1. *Fin-to-gate capacitance (C_{fg}):*
   C_{fg} is associated with electric field lines originating from the top surface of the silicon fin and terminating on the gate edge. The length of each electric field line is the perimeter of a quarter ellipse, which is approximated with the Euler approximation [27]:

$$\text{perimeter of ellipse} = 2\pi\sqrt{\frac{a^2+b^2}{2}}, \tag{7.39}$$

   where a is the length of the major axis and b is the length of the minor axis.
   As shown in Figure 7.25, each quarter ellipsoidal field line of C_{fg} has a major axis of l, and a minor axis of $T_{ox}+h$, where h ranges from 0 to H_{max}, and l ranges from 0 to L_{max}. We assume the ellipsoidal electric field lines do not go beyond L_{max} or H_{max}. Electric field lines that originate from the portion of the gate above H_{max} travel horizontally and land on the contact and become part of C_{cg1}, as we will show later. If H_g is less than H_{max}, then C_{cg1} becomes zero. H_{max} is extracted from TCAD simulations by varying H_g and finding the point where C_{cg1} diminishes to zero.
   Moreover, H_{max} is proportional to L_{ext} because increasing L_{ext} increases the distance between the gate and the contact and decreases the effect of the raised source/drain contact's inner surface on the fringe capacitance between the gate and the fin. Thus, we have

$$H_{max} = \frac{L_{ext}}{H_r}, \tag{7.40}$$

$$L_{max} = \frac{L_{ext}}{L_r}, \tag{7.41}$$

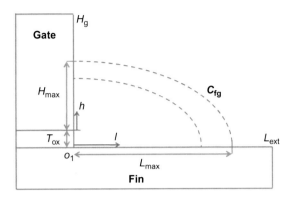

FIGURE 7.25

Electric field lines in the fin-to-gate region.

where $H_r = \text{HR0} + \text{HR1} \times \frac{H_g + T_{ox}}{H_c}$, and $L_r = \text{LRA} + \text{LRB} \times L_{ext} + \text{LRC}$ $\times \frac{H_c}{H_g + T_{ox}}$

If we limit h and l to H_{max} and L_{max}, the electric field lines in the gate-to-fin region have one axis of the ellipse varying from 0 to L_{max} and the other axis varying from T_{ox} to $H_{max} + T_{ox}$. To simplify the integration, the rates of change in the two axes are assumed to be proportional to each other. We may express h as a function of l:

$$h = \frac{H_{max}}{L_{max}} \cdot l = \frac{1}{H_r - L_r} \cdot l. \tag{7.42}$$

The length of the electric field line is calculated from Equations (7.39) to (7.42):

$$d(l) = \frac{1}{4}\left(2\pi \sqrt{\frac{\left(T_{ox} + \frac{1}{L_r H_r}\right)^2 + l^2}{2}}\right). \tag{7.43}$$

With use of Equations (7.38) and (7.43), the fin-to-gate capacitance per unit width becomes

$$C_{fg,sat} = \epsilon_{sp} \int_0^{L_{max}} \frac{1}{d(l)} dl = \epsilon_{sp} \frac{2\sqrt{2}}{\pi} \int_0^{L_{max}} \frac{1}{\sqrt{\left(T_{ox} + \frac{1}{H_r \cdot L_r}\right)^2 + l^2}} dl. \tag{7.44}$$

We integrate Equation (7.44) and introduce a constant fitting parameter CF1. $C_{fg,sat}$ becomes

$$C_{fg,sat} = \text{CF1} \cdot \frac{2\sqrt{2}\epsilon_{sp}r}{\pi\sqrt{r^2 + 1}} \cdot$$
$$\ln\left(\frac{\sqrt{(r^2 + 1)\left[(rT_{ox})^2 + 2rL_{max}T_{ox} + (r^2 + 1)L_{max}^2\right]} + rT_{ox} + (r^2 + 1)L_{max}}{rT_{ox}\left[1 + \sqrt{r^2 + 1}\right]}\right), \tag{7.45}$$

where $r = H_r \cdot L_r$.

Equation (7.45) is calculated under the assumption that gate height (H_g) is larger than H_{max}, and as a result C_{fg} does not change as we increase H_g. We denote this situation as the saturation condition. On the other hand, if H_g is less than H_{max}, the entire gate sidewall belongs to the fin-to-gate region. In this case C_{fg} is, to first order, a logarithmic function of H_g. We denote this as the log condition. $C_{fg,log}$ per unit length can be expressed as

$$C_{fg,log} = \epsilon_{sp}$$

$$\left[\frac{2\sqrt{2}}{\pi\sqrt{k+1}} \cdot \ln \left(\frac{\sqrt{k+1}\sqrt{(T_{ox} + H_g)^2 + (k \cdot H_g)^2} + T_{ox} + (k+1)H_g}{T_{ox}\left[(k+2) + \sqrt{(k+2)^2 + 1}\right]} \right) + WL \right],$$

(7.46)

where $k = \frac{L_{max}}{H_g + T_{ox}}$, and WL is a fitting parameter.

A smoothing function with a fitting parameter DELTA (δ) is used to describe the transition from $C_{fg,log}$ to $C_{fg,sat}$. The final C_{fg} is expressed as

$$C_{fg} = H_{fin} \left(C_{fg,sat} - \frac{(C_{fg,sat} - C_{fg,log} - \delta) + \sqrt{(C_{fg,sat} - C_{fg,log} - \delta)^2 + 4\delta C_{fg,sat}}}{2} \right).$$

(7.47)

2. *Source/drain contact to gate capacitance (C_{cg}):*
   The contact-to-gate capacitance, C_{cg}, describes the capacitance between the gate and the epitaxially grown source/drain contact. The contact is assumed to be rectangular, as shown in Figure 7.24. C_{cg} includes C_{cg1}, C_{cg2}, and C_{cg3}. Each component denotes the capacitance originating from different surfaces of the gate/contact. In addition, we introduce a fitting parameter C0CG to provide constant offset. The total C_{cg} can be expressed as

$$C_{cg} = H_{fin} \cdot (C_{cg1} + C_{cg2} + C_{cg3} + \epsilon_{sp} \cdot C0CG).$$

(7.48)

In the following sections, expressions for each capacitance are developed under the assumption that gate height is larger than the contact height ($H_g + T_{ox} > H_c$).

Note that of the three C_{cg} components, only C_{cg1} and C_{cg2} are implemented in BSIM-CMG because in practice, the region which C_{cg3} covers usually falls into the scope of adjacent structures, and the corresponding capacitances need not be included.

a. C_{cg1}:
   C_{cg1} is a simple parallel plate capacitance between the gate and the contact, as illustrated in Figure 7.24. Using Equation (7.38), we derive the capacitance per unit height:

$$C_{cg1} = \epsilon_{sp}\frac{H_g - H_{max}}{L_{ext}} \qquad\qquad \text{if } H_g \geq H_{max}.$$

(7.49)

As mentioned earlier, if $H_g < H_{max}$ then the entire inner sidewall of the gate belongs to the fin-to-gate region. This also means

$$C_{cg1} = 0 \qquad\qquad \text{if } H_g < H_{max}. \tag{7.50}$$

A smoothing function combines Equations (7.49) and (7.50) into a continuous equation:

$$C_{cg1} = \frac{1}{\text{CNON}} \cdot \ln\left[1 + \exp\left(\text{CNON} \cdot \epsilon_{sp}\frac{H_g - H_{max}}{L_{ext}}\right)\right], \tag{7.51}$$

where CNON ensures the exponential term is comparable to 1 when H_g is close to H_{max}.

However, Equation (7.51) assumes C_{cg1} spans the entire height of the gate above H_{max}. If the contact is not tall enough, this will become physically incorrect. Therefore, the final expression for C_{cg1} as implemented in BSIM-CMG is

$$C_{cg1} = \frac{1}{\text{CNON}} \cdot \ln\left[1 + \exp\left(\text{CNON} \cdot \epsilon_{sp}\frac{\min(H_c, H_g + T_{ox}) - H_{max}}{L_{ext}}\right)\right]. \tag{7.52}$$

Note that H_g in Equation (7.51) is replaced with $\min(H_c, H_g + T_{ox})$.

b. C_{cg2}:

The electric field lines of C_{cg2} originate from the gate sidewall, travel a distance L_{ext} horizontally, and then follow a quarter circle until they terminate on top of the contact (Figure 7.24). The quarter circle has a radius of r centered at the corner of the contact (marked as O_2 in Figure 7.24). Therefore, the expression for the length of the electric field lines is

$$d = L_{ext} + \frac{2\pi r}{4}, \tag{7.53}$$

where the radius, r, varies from 0 to R:

$$R = \frac{1}{2}\Delta H \cdot \frac{H_c}{H_g + T_{ox}} = \frac{H_c}{2}\left|1 - \frac{H_c}{H_g + T_{ox}}\right|, \tag{7.54}$$

$$\Delta H = |H_g + T_{ox} - H_c|.$$

ΔH is the difference between the gate height plus T_{ox} and the contact height. The capacitance in the region beyond R belongs to C_{cg3}. We perform integration from 0 to R to obtain C_{cg2} per unit height:

$$C_{cg2} = \epsilon_{sp} \int_0^R \frac{1}{L_{ext} + \frac{2\pi r}{4}} dr = \frac{2\epsilon_{sp}}{\pi}\ln\left(\frac{L_{ext} + 0.5\pi R}{L_{ext}}\right). \tag{7.55}$$

c. C_{cg3}:

C_{cg3} describes the capacitance with electric field lines originating from the top of the gate and terminating on the top of the contact. This is modeled by two capacitances in series: C_{cg3a} and C_{cg3b}. C_{cg3b} is a simple parallel plate capacitance characterized by the area of the contact surface and the horizontal distance:

$$C_{cg3b} = \epsilon_{sp} \frac{L_c - R}{\Delta H}. \tag{7.56}$$

C_{cg3a} is characterized by an equation similar to that for C_{fg} except with semicircular electric fields with diameters ranging from L_{ext} to $L_{ext} + L_c + L_g/2$. Only half of the gate length is considered because the other half would contribute to the same parasitic capacitance on the other side of the FinFET. Equation (7.57) results from a similar derivation:

$$C_{cg3a} = \frac{2\epsilon_{sp}}{\pi\left[(L_c - R) + L_g/2\right]} \ln\left(\frac{L_{ext} + L_c + L_g/2}{L_{ext} + R}\right). \tag{7.57}$$

The equation for C_{cg3a} is derived assuming that the gate is thicker than the contact epitaxy. If the opposite is true, then the values for $L_g/2$ and L_c are swapped, resulting in the same equation. The total C_{cg3} is described by the series combination of the two capacitances:

$$C_{cg3} = \frac{1}{\frac{1}{C_{cg3a}} + \frac{1}{C_{cg3b}}}. \tag{7.58}$$

As mentioned earlier in this section, C_{cg3} is not implemented in BSIM-CMG. Instead, capacitance in the region covered by C_{cg3} is modeled outside of BSIM-CMG.

d. Generalization to contacts taller than the gate:
Special consideration for the corner case where $H_g + T_{ox} < H_c$ is required. In FinFETs, the gate is usually taller than the raised source/drain contact in the vertical direction so we do not encounter this corner case. However, when we use the same model to describe parasitic capacitance on the side of the fin, H_g becomes the gate width on the side of the fin, and H_c becomes the contact width over the original fin. In such a case, $H_g + T_{ox}$ can be less than H_c for a single-fin FinFET or for the last fin in a multifin FinFET. Several simple changes to the C_{cg} model are required to cover such a case. First, the width of the parallel plate capacitance in C_{cg1} becomes $(H_g + T_{ox}) - (H_{max} + T_{ox})$. A more general expression for C_{cg1} is

$$C_{cg1} = \frac{1}{\text{CNON}} \cdot \ln\left[1 + \exp\left(\text{CNON} \cdot \epsilon_{sp} \frac{\min(H_g + T_{ox}, H_c) - (H_{max} + T_{ox})}{L_{ext}}\right)\right]. \tag{7.59}$$

Second, for C_{cg2}, a more general expression for R is

$$R = \frac{1}{2}\Delta H \cdot \min\left(\frac{H_g + T_{ox}}{H_c}, \frac{H_c}{H_g + T_{ox}}\right). \tag{7.60}$$

C_{cg3} does require minimal modification for the different geometry. All instances of L_c and $L_g/2$ are swapped with each other. C_{cg3a}, the semicircular portion, remains unchanged because the variables L_c and $L_g/2$ are commutative. The use of the absolute value in Equation (7.55) also allows ΔH to be general. The term $L_c - R$ in the numerator of

Equation (7.56) is replaced by $\frac{L_g}{2} - R$. In general, L_c in Equation (7.56) can be replaced by a variable L defined as

$$
\begin{cases}
L_c & \text{if } H_g + T_{ox} > H_c, \\
L_g & \text{if } H_g + T_{ox} < H_c.
\end{cases}
\tag{7.61}
$$

Equation (7.58) is still appropriate for C_{cg3}.

7.7 MODELING OF FinFET FRINGE CAPACITANCE IN THREE DIMENSIONS: CGEOMOD = 2

In the previous section, we developed a model for the region between the gate and the raised source/drain contact. In three-dimensional FinFETs (Figure 7.26) the situation is more complex. To simplify the problem, we separate the three-dimensional capacitances into a top component (C_{top}), two side components (C_{side}), and two corner components (C_{corner}), as illustrated in Figure 7.27.

The top component is exactly the two-dimensional capacitance model multiplied by the fin width (T_{fin}). The side component also uses the two-dimensional formulation, but with H_c replaced by T_c, and H_g replaced by W_g (see Figure 7.26 for symbol definitions), and multiplied by the fin height (H_{fin}). C_{corner} is modeled as a simple parallel plate capacitor with the two plates separated by a distance of L_{ext}:

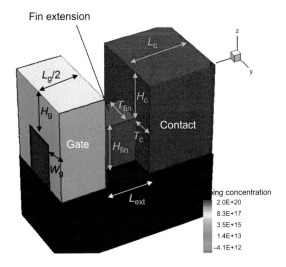

FIGURE 7.26

Structure of a single fin in a three-dimensional FinFET for TCAD simulations and symbol definitions.

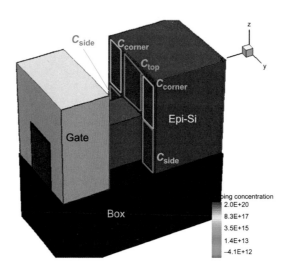

FIGURE 7.27

Composition of simpler capacitances to sum up to the whole parasitic capacitance.

$$C_{\text{corner}} = \epsilon_{\text{ox}} \frac{\min(T_{\text{c}}, W_{\text{g}} + T_{\text{ox}}) \cdot H_{\text{c}}}{L_{\text{ext}}}. \tag{7.62}$$

The total capacitance for a multifin FinFET with NFIN fins is modeled as

$$C_{\text{f}} = \text{NFIN} \times (2C_{\text{corner}} + 2C_{\text{side}} + C_{\text{top}}). \tag{7.63}$$

7.8 PARASITIC CAPACITANCE MODEL VERIFICATION

The analytical model for parasitic capacitance in FinFETs is developed with many degrees of freedom to allow a variety of device architectures to be modeled. Inherent to the model assumptions, there are a few fitting parameters whose value needs to be determined. We determine these values by comparison with numerical simulations (TCAD). Figures 7.28–7.30 verify the parasitic capacitance of each region by comparison with two-dimensional TCAD simulations. Figures 7.31 and 7.32 verify the total gate parasitic capacitance by comparison with three-dimensional TCAD simulations. The fitting parameter values for the best match with the TCAD simulations are listed in Table 7.5. All parameters are kept unchanged for these verifications, except for C0CG.

In Figure 7.28, the fin-to-gate capacitance (C_{fg}) per unit thickness is plotted versus H_{g} for different values of L_{ext}. As expected from the model, the saturation and log regions are shown: at first, C_{fg} grows logarithmically with H_{g} because $H_{\text{g}} + T_{\text{ox}}$ is still smaller than H_{max}. Then, C_{fg} becomes nearly constant because it goes into the saturation region after $H_{\text{g}} + T_{\text{ox}} > H_{\text{max}}$. Figure 7.29 exhibits contact source/drain

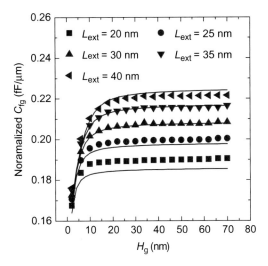

FIGURE 7.28

Normalized C_{fg} (fin-to-gate capacitance per unit thickness) versus H_g at different L_{ext}. The lines represent the model, and the symbols represent two-dimensional TCAD simulations.

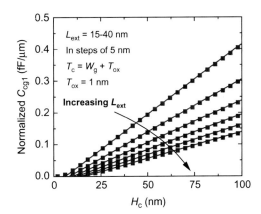

FIGURE 7.29

Normalized C_{cg1} (C_{cg1} per unit thickness) versus H_c at different L_{ext}. The lines represent the model, and the symbols represent two-dimensional TCAD simulations.

to gate capacitance versus H_c at different L_{ext}, where we keep $H_g + T_{ox} = H_c$. With $H_g + T_{ox} = H_c$ we force C_{cg2} to zero according to Figure 7.24. C_{cg3} is outside the simulation structure in this comparison. Figure 7.30 shows the verification of total C_{cg}. In this case we simulated a structure as depicted in Figure 7.24, where C_{cg2} and C_{cg3} are nonzero owing to uneven $H_g + T_{ox}$ and H_c.

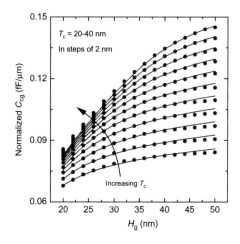

FIGURE 7.30

Normalized C_{cg} (C_{cg} per unit thickness) versus H_g at different H_c. COCG = 0.4. The lines represent the model, and the symbols represent two-dimensional TCAD simulations.

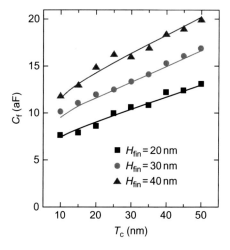

FIGURE 7.31

Total parasitic capacitance versus T_c at different H_{fin} for the middle-fin case. The lines represent the model, and the symbols represent three-dimensional TCAD simulations.

Figures 7.31 and 7.32 verify the total gate parasitic capacitance in the middle-fin and edge-fin cases from Figure 7.33, respectively. The value of C0CG is 0.47 in both situations. In the middle-fin case (Figure 7.31), the total parasitic capacitance is plotted versus T_c (equals $W_g + T_{ox}$) at $H_{fin} = 20, 30$, and 40 nm. In the edge-fin case (Figure 7.32), the total parasitic capacitance is plotted versus W_g at different T_c.

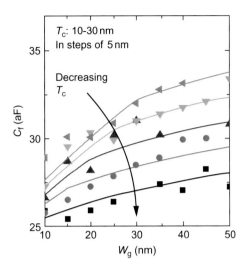

FIGURE 7.32

Total parasitic capacitance versus W_g at different T_c for the edge-fin case. The lines represent the model, and the symbols represent three-dimensional TCAD simulations.

Table 7.5 Extracted Parameters for the Compact Fringe Capacitance Model

Parameter Name	Value (TCAD Calibration)	Value (BSIM-CMG)[a]
LRA	1.05	1.05
LRB	0.0025	0.0
LRC	−0.01	0.0
WR0	2.3	2.3
WR1	0.2	0.2
HL	2.6	12.27
CF1	0.935	0.70 for a double-gate FinFET
		0.85 for a triple-gate FinFET
DELTA (δ)	1.2×10^{-12}	1.2×10^{-12}
C0CG[b]	0.47	0.47
CNON	1.7×10^{12}	1.7×10^{12}

[a]Values implemented in BSIM-CMG are a result of additional calibration, which gives slightly different values from those used in the verification shown in this chapter.

[b]C0CG is a fitting parameter which the user is allowed to tune to match data.

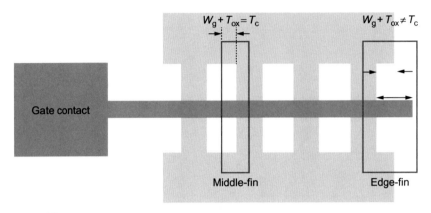

FIGURE 7.33

Top view of a four-fin FinFET with possible adaptations of the capacitance model to calculate the fringe capacitance in each section.

7.9 SUMMARY

In BSIM-CMG, we model both the parasitic resistance and the parasitic capacitances in FinFETs.

As far as parasitic resistance is concerned, we have developed a simple gate electrode resistance model, as well as a bias- and geometry-dependent parasitic source/drain resistance model. The gate electrode resistance model is capable of modeling single-sided contacts or double-sided contacts. As far as source and drain resistances are concerned, the geometry dependence of contact, spreading, and extension resistances is well modeled. The transmission-line-based contact resistance model predicts the dependence on epi height, fin pitch, contact length, cross-sectional shape of the epi, etc., and models various different contact silicidation schemes. Spreading resistance and extension resistance expressions were developed. The extension resistance exhibits bias dependence owing to the fringe field from the gate. The model was validated against TCAD simulations. The extension resistance model captures the gate bias dependence of source/drain resistance very well even for an underlapped device where the resistance has strong bias dependency. Resistance breakdown analysis shows that the spreading resistance is negligibly small, whereas extension and contact resistances dominate.

A parasitic gate capacitance model was developed for a two-dimensional slice of the FinFET structure and later extended to the complete three-dimensional structure using symmetry. It is composed of a gate-to-fin region (C_{fg}) that is characterized by elliptical field lines from the gate to the fin, and a gate-to-contact region (C_{cg}), further separated into the various possible surfaces on the contact or gate for the flux to pass through. C_{cg1} is essentially a parallel plate capacitance that levels off to zero at low gate or contact thicknesses. C_{cg2} has field lines emanating from the side of the gate to the top of the contact, which is characterized by a straight line plus a quarter-circle. C_{cg3} has field lines emanating from the top of the gate to the top of the contact. Its electric field lines can be modeled as a large half-circle plus straight line. Also, a parallel plate capacitance is used to describe C_{corner}. The total capacitance was verified with two-dimensional and three-dimensional TCAD simulations and was found to be accurate for a wide range of dimensions. The fitting parameters used were found to be suitable for the wide range of dimensions tested.

REFERENCES

[1] K. Mistry, et al., A 45 nm logic technology with high-k + metal gate transistors, strained silicon, 9 Cu interconnect layers, 193 nm dry patterning, and 100% Pb-free packaging, in: International Electron Devices Meeting Technical Digest, 2007, pp. 247–250.

[2] International Roadmap Committee, Process integration, devices, and structures, in: International Technology Roadmap for Semiconductors 2012 Update, 2012, Available: http://www.itrs.net/ [Online].

[3] W. Wu, M. Chan, Gate resistance modeling of multifin MOS devices, IEEE Electron Dev. Lett. 27 (1) (2006) 68–70.

[4] A. Dixit, A. Kottantharayil, N. Collaert, M. Goodwin, M. Jurczak, K. De Meyer, Analysis of the parasitic S/D resistance in multiple-gate FETs, IEEE Trans. Electron Devices 52 (6) (2005) 1132–1140.

[5] D. Tekleab, P. Zeitzoff, Modeling and analysis of parasitic resistance in double gate FinFETs, in: Proceedings of the IEEE International SOI Conference, 2008, pp. 51–52.

[6] J. Kedzierski, M. Ieong, E. Nowak, T.S. Kanarsky, Y. Zhang, R. Roy, D. Boyd, D. Fried, H.-S.P. Wong, Extension and source/drain design for high-performance FinFET devices, IEEE Trans. Electron Dev. 50 (4) (2003) 952–958.

[7] H. Shang, et al., Investigation of FinFET devices for 32 nm technologies and beyond, in: Symposium on VLSI Technology Digest of Papers, 2006, pp. 54–55.

[8] R.S. Sheony, K.C. Saraswat, Optimization of extrinsic source/drain resistance in ultrathin body double-gate FETs, IEEE Trans. Nanotechnol. 2 (4) (2003) 265–270.

[9] V. Trivedi, J.G. Fossum, M.M. Chowdhury, Nanoscale FinFETs with gate-source/drain underlap, IEEE Trans. Electron Devices 52 (1) (2005) 56–62.

[10] Sentaurus Device, Synopsys, Inc., September 2008.

[11] G. Masetti, M. Severi, S. Solmi, Modeling of carrier mobility against carrier concentration in arsenic-, phosphorus-, and boron-doped silicon, IEEE Trans. Electron Devices 30 (7) (1983) 764–769.

[12] H.H. Berger, Model for contacts to planar devices, Solid-State Electron. 15 (1972) 145–158.

[13] BSIM4 Model, Department of Electrical Engineering and Computer Science, UC Berkeley, Available: http://www-device.eecs.berkeley.edu/~bsim3/bsim4.html [Online].

[14] Sentaurus Structure Editor, Synopsys, Inc., September 2008.

[15] Noffset 3D, Synopsys, Inc., September 2008.

[16] H. Kawasaki, et al., Demonstration of highly scaled FinFET SRAM cells with high-K metal gate and investigation of characteristic variability for the 32 nm node and beyond, in: International Electron Devices Meeting Technical Digest, 2008, pp. 237-240.

[17] C. Lombardi, S. Manzini, A. Saporito, M. Vanzi, A physically based mobility model for numerical simulation of nonplanar devices, IEEE Trans. Comput. Aided Des. 7(11) (1988) 1164-1171.

[18] C. Canali, G. Majni, R. Minder, G. Ottaviani, Electron and hole drift velocity measurements in silicon and their empirical relation to electric field and temperature, IEEE Trans. Electron Devices 22(11) (1975) 1045-1047.

[19] R. Vega, T.-J. King, Three-dimensional FinFET source/drain and contact design optimization study, IEEE Trans. Electron Devices 56(7) (2009) 1483-1492.

[20] International Roadmap Committee, Interconnect, in: International Technology Roadmap for Semiconductors 2010 Update, 2010, Available: http://www.itrs.net/ [Online].

[21] N. Stavitski, M.J.H. van Dal, A. Lauwers, C. Vrancken, A.Y. Kovalgin, R.A.M. Wolters, Systematic TLM measurements of NiSi and PtSi specific contact resistance to n- and p-type Si in a broad doping range, IEEE Electron Dev. Lett. 29(4) (2008) 378-381.

[22] R.S. Shenoy, K.C. Saraswat, Optimization of extrinsic source/drain resistance in ultrathin body double-gate FETs, IEEE Trans. Nanotechnol. 2(4) (2003) 265-270.

[23] V.P. Trivedi, J.G. Fossum, Quantum-mechanical effects on the threshold voltage of undoped double-gate MOSFETs, IEEE Electron Device Lett. 29 (8) (2005) 579–582.

[24] R.H. Dennard, F.H. Gaensslen, H.-N. Yu, V.L. Rideout, E. Bassous, A.R. Leblanc, Design of ion implanted MOSFET's with very small dimensions, IEEE J. Solid-State Circ. 9(5) (1974) 256-268.

[25] S. Venugopalan, N. Paydavosi, J. Duarte, D. Lu, C.-H. Lin, M. Dunga, S. Yao, T. Morshed, A. Niknejad, C. Hu, BSIM-CMG 107.0.0 Multi-Gate MOSFET Compact Model, Technical Manual, 2013.

[26] K. Suzuki, Parasitic capacitance of submicrometer MOSFET's, IEEE Trans. Electron Dev. 46(9) (1999) 1895-1900.

[27] S. Sykora (Ed.), Approximations of Ellipse Perimeters, Stan's Library, vol. I, Castano Primo, Italy, December 2005, Available: http://www.ebyte.it/library/docs/math05a/EllipsePerimeterApprox05.html [Online].

Noise

8

CHAPTER OUTLINE

8.1 Introduction 195
8.2 Thermal noise... 196
8.3 Flicker noise 198
8.4 Other noise components 201
8.5 Summary 201
References... 201

8.1 INTRODUCTION

From a microscopic point of view, carrier transport in electronic devices is a stochastic process. Electrons and holes move randomly at their thermal velocity ($\sqrt{3kT/m} \approx 10^7$ cm/s). The transport models and BSIM-CMG I-V expressions described in previous chapters model only the average current, which does not take into account such random motion of carriers. On the other hand, the stochastic nature of carrier movement gives rise to a time-dependent fluctuation in the current. This phenomenon is described by noise models. In this chapter, we will focus on the noise models in BSIM-CMG.

The physics of noise in FinFETs of fully depleted devices is similar to that in planar MOSFETs. For this reason, the model expressions well established for planar MOSFETs, such as those in the BSIM4 model [1], can be adopted with some modifications. Quantitative differences, such as different mobility owing to different channel surface orientation in FinFETs (mostly $\langle 110 \rangle$) versus planar devices ($\langle 100 \rangle$), are accounted for via the use of FinFET mobility parameter values.

Noise components include thermal noise associated with the channel, the source, the drain, and the substrate, flicker or $1/f$ noise, and shot noise, as we will describe in the following sections.

8.2 THERMAL NOISE

Thermal noise (*Nyquist noise* or *Johnson noise*) associated with a conductive (or resistive) element is given in terms of the noise spectral density:

$$\frac{S}{\Delta f} = 4kTG, \tag{8.1}$$

where G is the conductance of the element. For a MOSFET operating in the linear region, the channel conductance can be expressed in the long-channel limit as

$$G = \frac{I_{ds}}{V_{ds}} = \mu_{eff} \frac{W}{L_{eff}} Q_{inv} = \frac{\mu_{eff} q_{inv}}{L_{eff}^2}, \tag{8.2}$$

where Q_{inv} denotes the inversion charge density per unit area, and q_{inv} is the total inversion charge in the device channel. The above expression is well established for long-channel MOSFETs.

The total resistance is the sum of the channel resistance and external source and drain resistances:

$$\frac{1}{G} = R_{dsi} + \frac{L_{eff}^2}{\mu_{eff} q_{inv}}. \tag{8.3}$$

Substituting Equation (8.3) into Equation (8.1), we have

$$S_{id} = \frac{4kT \mu_{eff} q_{inv}}{\mu_{eff} q_{inv} R_{dsi} + L_{eff}^2}. \tag{8.4}$$

where we denoted the drain noise spectral density as S_{id}. The above expression is identical to the charge-based model in BSIM4 (selected by setting tnoiMod $= 0$ in BSIM4).

For short-channel devices, drain noise is usually larger than what the long-channel theory predicts. The noise increase for short-channel devices is typically accounted for with a multiplication factor γ. In BSIM-CMG, γ is represented by the model parameter NTNOI. The final thermal noise expression including NTNOI becomes

$$S_{id} = NTNOI \cdot \frac{4kT \mu_{eff} q_{inv}}{\mu_{eff} q_{inv} R_{dsi} + L_{eff}^2}. \tag{8.5}$$

Although Equation (8.5) is derived for MOSFET operation in the linear region, the same expression is applicable to the saturation region as well. This is because the thermal noise in the saturation region coincides with that in the linear region near the onset of saturation (Figure 8.1).

Besides noise from the drain to the source (S_{id}), thermal noise due to carrier fluctuation in the channel can be coupled to the gate as well through the gate dielectric. This is known as *induced gate noise*. The theory for induced gate noise modeling is well established [2–6]. Induced gate noise is proportional to $\omega^2 L_{eff}^2$, where ω is 2π times the frequency of operation. Therefore, induced gate noise is important for longer-channel devices or at very high frequency (Figure 8.2). A model for induced gate noise was developed and available for implementation in BSIM-CMG [2].

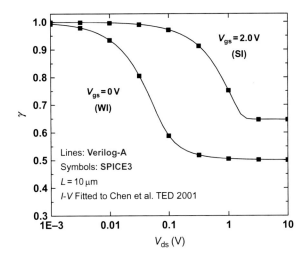

FIGURE 8.1

Simulated γ versus drain voltage in the weak ($V_{gs} = 0$) and strong ($V_{gs} = 2.0$) inversion regions for a long-channel planar MOSFET. In weak inversion, drain noise coincides with $2qI_d$ [2].

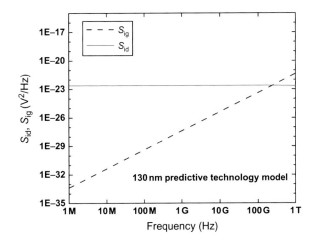

FIGURE 8.2

Evaluation of drain noise (S_{id}) and induced gate noise (S_{ig}) versus frequency for the 130 nm BSIM4-based predictive technology model [7] ($L = 130$ nm, tnoiMod = 1, fnoiMod = 0). S_{ig} is much smaller than S_{id} until the frequency of operation is larger than 100 GHz.

8.3 FLICKER NOISE

Besides thermal noise, *flicker noise*, or $1/f$ noise, is another major source of MOS drain current noise, especially at low frequency. Its origin is found to be related to traps in the gate dielectric. There are two theories that explain the physical mechanism of $1/f$ noise: *carrier number fluctuation theory* and *mobility fluctuation theory*. In the carrier number fluctuation theory, the flicker noise is attributed to the random trapping and de-trapping processes of charges in the oxide traps near the Si-SiO$_2$ interface. The charge fluctuation results in fluctuation in the channel carrier density, which in turn modulates the drain current [8]. In the mobility fluctuation theory [9], on the other hand, the current level increases/decreases as a result of the fluctuation in bulk mobility.

The *unified model* [10, 11], which accounts for both mechanisms, is adopted in BSIM-CMG. The same model was used in BSIM4 [1]. In this section, we describe the development of the unified noise model.

The drain current in a MOSFET can be expressed as

$$I_d = W\mu q N \varepsilon_x, \tag{8.6}$$

where ε_x is the lateral electric field. By taking the derivative and rearranging terms, we have

$$\frac{1}{I_d}\frac{\delta I_d}{\delta N_t} = \underbrace{\frac{1}{N}\frac{\delta N}{\delta N_t}}_{\text{carrier number fluctuation}} \pm \underbrace{\frac{1}{\mu}\frac{\delta \mu}{\delta N_t}}_{\text{mobility fluctuation}}, \tag{8.7}$$

where we mathematically separate the drain current variation in terms of the two underlying physical mechanisms. In Equation (8.7), the channel carrier fluctuation, δN, and the trap number fluctuation, δN_t are closely related via capacitive coupling:

$$R = \frac{\delta N}{\delta N_t} = -\frac{C_{inv}}{C_{ox} + C_{inv} + \text{CIT}} = -\frac{N}{N + N^*}, \tag{8.8}$$

where

$$N^* = \frac{kT}{q^2}(C_{ox} + \text{CIT}) \tag{8.9}$$

and CIT is a BSIM-CMG model parameter for the interface trap coupling capacitance. The mobility term in Equation (8.7) can be expressed using the Matthiessen rule (using n-channel device as an example):

$$\frac{1}{\mu} = \frac{1}{\mu_n} + \frac{1}{\mu_{ox}} = \frac{1}{\mu_n} + \alpha N_t. \tag{8.10}$$

where μ_n is the electron mobility in the channel and α is the scattering parameter for the influence of oxide traps charges on mobility. Differentiating (8.10) with respect to N_t we have

$$\frac{\delta \mu}{\delta N_t} = -\alpha \mu^2. \tag{8.11}$$

Substituting Equations (8.8) and (8.11) into Equation (8.7), we obtain

$$\frac{\delta I_d}{I_d} = \left(\frac{1}{N} \cdot R \pm \alpha\mu\right) \delta N_t. \tag{8.12}$$

We may express the local current fluctuation in terms of trap density fluctuation:

$$S_{\Delta I_d} = \left(\frac{R}{N} \pm \alpha\mu\right)^2 I_d^2 \cdot S_{\Delta N_t}, \tag{8.13}$$

where the trap density fluctuation is given by

$$S_{\Delta N_t} = \int_{E_v}^{E_c} \int_0^W \int_0^{T_{ox}} 4N_t f_t (1 - f_t) \frac{\tau}{1 + \omega^2\tau^2} dzdyd\varepsilon$$

$$= N_t(E_{fn}) \cdot \frac{kTW\Delta x}{\gamma f}, \tag{8.14}$$

where the Lorentzian noise spectrum associated with each individual trap is integrated. In the above expression, γ is the exponential coefficient for the position dependence of trapping time constant τ, and τ is an exponential function of position in the gate oxide:

$$\tau = \tau_0(E)e^{\gamma z}, \tag{8.15}$$

The integration has produced a $1/f$ term in Equation (8.14). Details of the integration can be found in [11]. The local current fluctuation is integrated to obtain the noise spectral density at the drain terminal:

$$S_{id} = \frac{1}{L^2} \int_0^L \frac{kTI_d^2}{\gamma f} N_t(E_{fn}) \left(\frac{R}{N} \pm \alpha\mu\right)^2 dx$$

$$= \frac{1}{L^2} \int_0^L \frac{kTq\mu}{\gamma f} N_t(E_{fn}) \left(1 \pm \frac{\alpha\mu N}{R}\right)^2 \frac{R^2}{N} dV. \tag{8.16}$$

We express the trap density as a parabolic function of the carrier density with three parameters—A, B, and C:

$$N_t^*(E_{fn}) = N_t(E_{fn}) \left(1 \pm \frac{\alpha\mu N}{R}\right)^2$$

$$= A + BN + CN^2. \tag{8.17}$$

We then carry out the integration in Equation (8.16) and obtain the drain noise spectral density expression for the linear (triode) region:

$$S_{id} = \frac{kTq^2 I_d\mu}{\alpha\gamma f L^2 C_{ox}} \left[A \ln\left(\frac{N_0 + N^*}{N_L + N^*}\right) + B(N_0 - N_L) + \frac{1}{2}C\left(N_0^2 - N_L^2\right)\right]. \tag{8.18}$$

Here N_0 and N_L are the carrier densities at the source and the drain, respectively:

$$N_0 = \frac{C_{ox} \cdot q_{is}}{q}, \tag{8.19}$$

$$N_L = \frac{C_{ox} \cdot q_{id}}{q}. \tag{8.20}$$

The above expression for S_{id} is for noise in the triode region. For the pinch-off region, we simply substitute Equation (8.17) into Equation (8.16) and evaluate the integrand only at $x = L$ and substitute dx with ΔL_{clm}, the channel length modulation distance. The expression becomes

$$S_{id} = \Delta L_{clm} \cdot \frac{kTI_d^2}{\gamma f W L^2} \frac{A + BN_L + CN_L^2}{(N_L + N^*)^2}, \tag{8.21}$$

where

$$\Delta L_{clm} = l \cdot \ln\left[\frac{1}{E_{sat,noi}} \cdot \left(\frac{V_{ds} - V_{dseff}}{l} + EM\right)\right]. \tag{8.22}$$

where EM is a BSIM-CMG model parameter for channel length modulation and $E_{sat,noi}$ is the saturation electric field for noise modeling. The final unified flicker noise expression for the strong-inversion region is implemented in BSIM-CMG by adding Equations (8.18) and (8.21):

$$S_{si} = \frac{kTq^2 \mu_{eff} I_{ds}}{C_{ox} L_{eff,noi}^2 \cdot f^{EF} \cdot 10^{10}} \cdot FN1 + \frac{kTI_{ds}^2 \Delta L_{clm}}{W_{eff} \cdot NFIN_{total} \cdot L_{eff,noi}^2 \cdot f^{EF} \cdot 10^{10}} \cdot FN2, \tag{8.23}$$

where $\gamma = 10^{10}$ is found via experiments. Although flicker noise is proportional to $1/f$ in theory, the adjustable parameter EF allows the model slope to deviate slightly from unity. FN1 and FN2 are given by,

$$FN1 = NOIA \cdot \ln\left(\frac{N_0 + N^*}{N_L + N^*}\right) + NOIB \cdot (N_0 - N_L) + \frac{NOIC}{2}(N_0^2 - N_L^2), \tag{8.24}$$

$$FN2 = \frac{NOIA + NOIB \cdot N_L + NOIC \cdot N_L^2}{(N_L + N^*)^2}, \tag{8.25}$$

where NOIA, NOIB, and NOIC are nothing but A, B, and C. For the weak-inversion region, the same S_{id} expression can be used except the Q-V relationship is different. The subthreshold noise expression can be found by evaluating Equation (8.16) with

$$\frac{dN}{dV} = -\beta. \tag{8.26}$$

The weak-inversion S_{id} is

$$S_{wi} = \frac{NOIA \cdot kT \cdot I_{ds}^2}{W_{eff} \cdot NFIN_{total} \cdot L_{eff,noi} \cdot f^{EF} \cdot 10^{10} \cdot N^{*2}}. \tag{8.27}$$

The weak-inversion (Equation 8.27) and strong-inversion (Equation 8.23) region formula are combined with the formula of harmonic means:

$$S_{id,flicker} = \frac{S_{wi}S_{si}}{S_{wi} + S_{si}}. \tag{8.28}$$

The above expression is implemented in BSIM-CMG and other BSIM models for flicker noise calculations.

8.4 OTHER NOISE COMPONENTS

Parasitic resistors in a MOS device also contribute to noise. For RDSMOD = 0, the source and drain resistance thermal noise is considered in the channel thermal noise expression, (Equation 8.5). On the other hand, for RDSMOD = 1, the source and drain resistances are modeled as resistor elements. In this case we model thermal noise associated with R_{source} and R_{drain} using Equation (8.1):

$$\frac{\overline{i_{RS}^2}}{\Delta f} = 4kT \cdot \frac{1}{R_{source}}, \tag{8.29}$$

$$\frac{\overline{i_{RD}^2}}{\Delta f} = 4kT \cdot \frac{1}{R_{drain}}, \tag{8.30}$$

where $\overline{i_{RS}^2}$ and $\overline{i_{RD}^2}$ are noise current sources in parallel to R_{source} and R_{drain}.

Similarly, if the gate electrode resistance is modeled as a resistor element (RGATEMOD = 1), we have

$$\frac{\overline{i_{RG}^2}}{\Delta f} = 4kT \cdot \frac{1}{R_{geltd}}. \tag{8.31}$$

The noise associated with quantum mechanical tunneling is shot noise. Therefore, each gate current component has a corresponding shot noise component given by $2qI$, where I is the tunneling current:

$$\overline{i_{gs}^2} = 2q(I_{gcs} + I_{gs}), \tag{8.32}$$

$$\overline{i_{gd}^2} = 2q(I_{gcd} + I_{gd}), \tag{8.33}$$

$$\overline{i_{gb}^2} = 2qI_{gbinv}. \tag{8.34}$$

8.5 SUMMARY

We have derived the thermal noise expression for BSIM-CMG, in which the drain noise is expressed as a function of the total inversion charge. Flicker noise is modeled based on the unified model, which accounts for both carrier number fluctuation and mobility fluctuation. Shot noise generated from gate tunneling current was modeled as well. Finally, thermal noise associated with each resistor element is properly accounted for in BSIM-CMG.

REFERENCES

[1] BSIM4 MOSFET Model User's Manual, Available: http://www-device.eecs.berkeley.edu/bsim/ [Online].
[2] D. Lu, Compact Models for Future Generation CMOS, Ph.D. Dissertation, UC Berkeley, 2011.

[3] A. Van der Ziel, Gate noise in field effect transistors at moderate high frequencies, Proc. IEEE 51 (3) (1963) 461–467.

[4] J.C.J. Paasschens, A.J. Scholten, R. van Langevelde, Generalizations of the Klaassen-Prins equation for calculating the noise of semiconductor devices, IEEE Trans. Electron Devices 52 (11) (2005) 2463–2472.

[5] A.J. Scholten, L.F. Tiemeijer, R. van Langevelde, R.J. Havens, A.T.A. Zegers-van Duijnhoven, V.C. Venezia, Noise modeling for RF CMOS circuit simulation, IEEE Trans. Electron Devices 50 (3) (2003) 618–632.

[6] J.-S. Goo, W. Liu, C. Chang-Hoon, K.R. Green, Z. Yu, T.H. Lee, R.W. Dutton, The equivalence of van der Ziel and BSIM4 models in modeling the induced gate noise of MOSFETs, in: International Electron Devices Meeting Technical Digest, 2000, pp. 811–814.

[7] Predictive Technology Model (PTM), School of Electrical, Computer, and Electrical Engineering, Arizona State University, Available: http://ptm.asu.edu/ [Online].

[8] S. Christensson, I. Lundstrom, C. Svensson, Low frequency noise in MOS transistors—I theory, Solid-State Electron. 11 (1968) 797.

[9] L.K.J. Vandamme, Model for $1/f$ noise in MOS transistors biased in the linear region, Solid-State Electron. 23 (1980) 317.

[10] K.K. Hung, P. Ko, C. Hu, Y. Cheng, A unified model for the flicker noise in metal-oxide-semiconductor field-effect transistors, IEEE Trans. Electron Devices 37 (3) (1990) 654–665.

[11] K.K. Hung, P.K. Ko, C. Hu, Y.C. Cheng, A physics-based MOSFET noise model for circuit simulations, IEEE Trans. Electron Devices 37 (5) (1990) 1323–1333.

Junction diode *I-V* and *C-V* models

CHAPTER OUTLINE

9.1 Junction diode current model 205
  9.1.1 Reverse-bias additional leakage model 208
9.2 Junction diode charge/capacitance model 210
  9.2.1 Reverse-bias model 210
  9.2.2 Forward-bias model 213
References 216

As the channel length is scaled down, the traditional planar bulk MOSFET incurs an increased amount of undesired subsurface leakage current from the source to the drain. This region below the interface is controlled less by the gate and more by the source-drain electric fields. In planar transistors, halo implants below the source and drain regions were used to mitigate this current. With decreasing channel length, we saw the need for increasing the number of halo implants that almost merged together from either side. However with further scaling, an ever-higher halo implant dosage could not be used. The semiconductor industry moved away from planar transistors, creating the need for multigate transistors where the three-dimensional nature of the channel allowed better gate control over the channel. Today, FinFETs on bulk and silicon-on-insulator substrates have moved into mass production.

FinFETs, owing to their three-dimensional nature, are manufactured through a set of process steps that slightly differ from those used for the traditional bulk planar MOSFETs. FinFETs on a bulk substrate have a region just below the fin that is less controlled by the gate. An electric field from the source and drain regions would extend into the region below the fin, creating subsurface leakage (although lesser in magnitude compared with planar transistors of the same channel length). In order to weaken these fields, some amount of punch-through stop (PTS) implant (aka ground plane doping) is employed for bulk FinFETs in production [1]. Figure 9.1 illustrates a cross-section of a *p*-type FinFET showing the presence of this implant just below the actual fin region. This implant below the intrinsic fin region will diffuse laterally under the source/drain junction region. The use of a high-dose well

FinFET Modeling for IC Simulation and Design. http://dx.doi.org/10.1016/B978-0-12-420031-9.00009-9

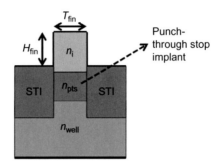

FIGURE 9.1

Cross-section of a bulk *p*-type FinFET showing a punch-through stop implant below the fin. Gate and oxide regions are not shown.

implant will lead to an increase of the junction tunneling current leakage component. Thus, the magnitude of doping in this region will have to be less than or equal to the magnitude of well doping. In Figure 9.2, the cross-section of a *p*-type FinFET under the source/drain region is illustrated; this leads to the creation of a double junction with two different *n*-type dopings $p^+|n_{pts}|n_{well}$ as a general case. When the reverse bias applied to this junction is increased (e.g., through increased positive drain voltage), the depletion region edge will traverse through the n_{pts} region and could enter the n_{well} region too. This leads to a deviation in the behavior of source/drain junction diodes from that of an ideal uniformly doped $p^+|n$ type of step junction diode observed in planar bulk MOSFETs. In terms of just the junction diode currents, the difference is imperceptible as the reverse-bias currents are rather very small in

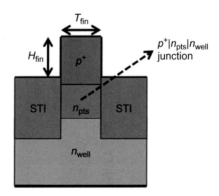

FIGURE 9.2

Punch-through stop implant below the fin diffuses laterally below the source/drain junction region, leading to a $p^+|n_{pts}|n_{well}$ type of junction. Gate and oxide regions are now shown.

magnitude. For this reason, the junction diode current model in BSIM-CMG follows a modeling philosophy similar to that for planar bulk MOSFETs (such as in the BSIM4 model). A comprehensive junction diode current model that captures reverse breakdown as well as the effect of finite series parasitic resistance has been adopted in BSIM-CMG. However, the behavior of reverse-bias junction diode capacitance has been observed to be markedly different from that in planar bulk MOSFETs. A new junction capacitance model has been developed to capture this process-induced subtlety in the bulk FinFET junction region. Compared with BSIM4, the forward-bias junction diode capacitance model has been improved as well by taking into consideration the accuracy of higher-order harmonic currents for RF CMOS IC designs.

9.1 JUNCTION DIODE CURRENT MODEL

From the Boltzmann transport model one can derive an analytic expression for the ideal junction current as follows:

$$I_{jn} = I_{jn0} \left(e^{\frac{qV_{jn}}{kT}} - 1 \right), \tag{9.1}$$

where V_{jn} is the voltage bias across the diode and I_{jn0} is the reverse-bias saturation current of the junction. In BSIM-CMG (similarly to compact models for planar bulk MOSFET), the source/drain-substrate junction diode regions are broken down into three distinct regions for the purpose of modeling—the junction bottom or the area component, the shallow trench isolation (STI) side junction or the perimeter component, and the MOSFET gate-edge component. These are captured through a distinct set of parameters for the source side and the drain side. In what follows, we will restrict our discussion to the source-side junction diode current. The drain-side junction diode current equations resemble the source-side junction current equations but with a different set of parameters. The different sets of parameters for the source side and the drain side allow us to capture any asymmetry in the source-side and drain-side junctions such as in a multifinger FinFET.

The reverse-bias saturation current is given by

$$I_{jns0} = \text{ASEJ} \cdot \text{JSS} + \text{PSEJ} \cdot \text{JSWS} + W_{eff} \cdot \text{NF} \cdot \text{NFIN} \cdot \text{JSWGS}, \tag{9.2}$$

where ASEJ is the bottom area and PSEJ is the STI side perimeter of the junction, and JSS, JSWS, and JSWGS are the bottom, STI side and gate-edge reverse-bias saturation current densities. Under reverse-bias conditions, the ideal diode current equation, Equation (9.1) predicts a constant reverse saturation current. However, real junction diodes exhibit junction current that tends to increase with the magnitude of the reverse bias (owing to various leakage currents, as discussed later) and after a certain reverse-bias voltage, an exponential increase in current due to a phenomenon known as breakdown. For large reverse-bias voltages, the depletion region of the junction experiences high electric fields that tend to accelerate electrons and holes, which in turn knock off more electrons from the lattice of the material, thus contributing to an avalanche of current. Under similar conditions of high field, in some highly doped junctions, electrons can jump from one side of the junction to the other through a quantum mechanical effect known as tunneling. This reverse breakdown current is captured in the BSIM-CMG model as follows:

$$I_{jns} = I_{jns0} \left(e^{\frac{qV_{es}}{NJS \cdot kT}} - 1 \right) \cdot F_{breakdown},$$
(9.3)

where the breakdown factor, $F_{breakdown}$, is empirically modeled as

$$F_{breakdown} = 1 + XJBVS \cdot e^{-\frac{q(BVS+V_{es})}{NJS \cdot kT}}.$$
(9.4)

In the above equations, V_{es} is the substrate to source terminal bias voltage. XJBVS, the breakdown coefficient, has a default value of 1. XJBVS can be conveniently set to zero if one does not want to capture breakdown through the model. BVS represents the source-to-substrate breakdown voltage. The nonideality factor NJS is used to capture any inadequacies in the junction diode current model with respect to a real diode such as additional leakage currents due to traps in the junction region (discussed later).

Under forward-bias conditions, the junction diode current equation, Equation (9.3), is a purely exponential function which could potentially cause convergence issues in a SPICE simulator environment. Realistic diodes are far from ideal, and the quasi-neutral regions of the diode present a finite series resistance, restricting the diode current behavior to a linear dependence on junction bias voltage. The situation under reverse breakdown conditions is similar. In BSIM-CMG, the effect of the series resistances is captured by making the junction diode currents transition smoothly to a linear behavior with respect to the voltage bias on the current exceeding and after it has exceeded a certain absolute value. In the model, for the forward-bias region, this value is set by the parameter IJTHSFWD, and for the reverse-bias region it is set through the parameter IJTHSREV. Hence, the junction diode current formulation can be divided into three regions—less than IJTHSREV, in between IJTHSREV and IJTHSFWD, and greater than IJTHSFWD (Figure 9.3). For the region in between IJTHSREV and IJTHSFWD, the diode current is described in Equation (9.3). Let us now consider the case where the diode current is greater than IJTHSFWD. The bias voltage, V_{jsmFwd}, at which this occurs can be found by substituting IJTHSFWD and V_{jsmFwd} into Equation (9.3). Regrouping similar terms, one obtains a quadratic equation as follows:

$$X^2 - BX - C = 0,$$
(9.5)

where

$$X = e^{\frac{q \cdot V_{jsmFwd}}{NJS \cot kT}},$$

$$B = 1 + \frac{IJTHSFWD}{I_{jns0}} - XJBVS \cdot e^{-\frac{q \cdot BVS}{NJS \cdot kT}},$$

$$C = XJBVS \cdot e^{-\frac{q \cdot BVS}{NJS \cdot kT}}.$$

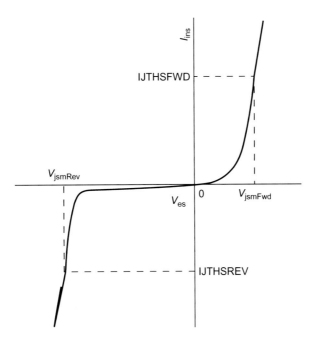

FIGURE 9.3

Junction diode current for the three regions as described in the BSIM-CMG model.

One can then solve for the value of V_{jsmFed} as

$$V_{jsmFwd} = \frac{NJS \cdot kT}{q} \log\left(\frac{B + \sqrt{B^2 + 4C}}{2}\right). \tag{9.6}$$

A first-order Taylor expansion of Equation (9.3) at $V_{es} = V_{jsmFwd}$ is then used to obtain the junction diode current expression for a diode junction current greater than IJTHSFWD, which is given as follows:

$$I_{jns} = IJTHSFWD + k_{slopeFwd} \cdot (V_{es} - V_{jsmFwd}) \tag{9.7}$$

where the slope factor $k_{slopeFwd}$ can be obtained from the first-order derivative of Equation (9.3) with respect to V_{es} at $V_{es} = V_{jsmFwd}$:

$$k_{slopeFwd} = \frac{qI_{jns0}}{NJS \cdot kT}\left(e^{\frac{qV_{jsmFwd}}{NJS \cdot kT}} + XJBVS \cdot e^{-\frac{BVS + V_{es}}{NJS \cdot kT}}\right). \tag{9.8}$$

For reverse bias, the diode current equation, Equation (9.3), captures reverse breakdown. However, for large reverse bias (negative V_{es}) the term $F_{breakdown}$ dominates and increases exponentially. The effect of series resistance is captured by making the diode junction current linear in a fashion similar to the way we dealt with forward-bias behavior. Under reverse breakdown, the diode junction current can be approximated as follows:

$$I_{\text{jns,rev}} = I_{\text{jns0}} F_{\text{breakdown}} \tag{9.9}$$

The bias voltage at which we transition to a linear behavior, V_{jsmRev}, can be found by substituting $I_{\text{jns,rev}} = \text{IJTHSREV}$ and $V_{\text{es}} = V_{\text{jsmRev}}$ into the above equation. We then arrive at

$$V_{\text{jsmRev}} = -\text{BVS} - \frac{\text{NJS} \cdot kT}{q} \log\left(\frac{\text{IJTHSREV} - I_{\text{jns0}}}{\text{XJBVS} \cdot I_{\text{jns0}}}\right). \tag{9.10}$$

With a first-order Taylor expansion of Equation (9.9) at $V_{\text{es}} = V_{\text{jsmRev}}$, the reverse-bias junction diode current for values less than IJTHSREV is given by

$$I_{\text{jns}} = \left(e^{\frac{qV_{\text{es}}}{\text{NJS} \cdot kT}} - 1\right)\left[\text{IJTHSREV} + k_{\text{slopeRev}}(V_{\text{es}} - V_{\text{jsmRev}})\right]. \tag{9.11}$$

Finally, the overall source-side diode junction current in the BSIM-CMG model is given as

$$I_{\text{jns}} = \begin{cases} \left(e^{\frac{qV_{\text{es}}}{\text{NJS} \cdot kT}} - 1\right)\left[\text{IJTHSREV} + k_{\text{slopeRev}}(V_{\text{es}} - V_{\text{jsmRev}})\right] & V_{\text{es}} < V_{\text{jsmRev}}, \\ I_{\text{jns0}}\left(e^{\frac{qV_{\text{es}}}{\text{NJS} \cdot kT}} - 1\right) \cdot F_{\text{breakdown}} & V_{\text{jsmRev}} < V_{\text{es}} < V_{\text{jsmFwd}}, \\ \text{IJTHSFWD} + k_{\text{slopeFwd}} \cdot (V_{\text{es}} - V_{\text{jsmFwd}}) & V_{\text{es}} > V_{\text{jsmFwd}}. \end{cases} \tag{9.12}$$

9.1.1 REVERSE-BIAS ADDITIONAL LEAKAGE MODEL

Junction diodes in MOSFETs are known to exhibit three different leakage current phenomena.

Firstly, the electron and hole generation-recombination trap centers are known to lead to Shockley-Reed-Hall (SRH) current. These trap centers are present owing to impurities and physical lattice defects in the silicon that are formed during the junction formation. Semiconductor fabrication steps such as annealing strive to keep the density of these trap centers below a certain limit. Of these traps, only the traps that fall into the depletion region of the junctions contribute to the leakage current. In the forward-bias mode of operation, these traps facilitate net electron-hole recombination, while under reverse bias there is net electron-hole generation. The built-in electric field in the depletion regions helps separate/recombine the electron-hole pair, leading to the SRH leakage current.

Heavily doped ($10^{19}\,\text{cm}^{-3}$) junctions also exhibit leakage current owing to band-to-band tunneling (BTBT). BTBT-based current is usually observed when the built-in electric field exceeds $10\,\text{MV/cm}$. This quantum mechanical process occurs without the aid of traps across a physically thin energy barrier region. This component of current is proportional to $E_{\text{dep,max}}^2 e^{\frac{1}{E_{\text{dep,max}}}}$, where $E_{\text{dep,max}}$ is the magnitude of the maximum built-in electric field in the depletion region. This electric field is higher under reverse bias than forward bias of operation and hence the BTBT current is higher as well. However, such high amounts of doping are usually avoided in carefully engineered modern CMOS devices to keep this leakage component well under control. For this reason, we will not aim to capture this component in our model.

Finally the trap centers discussed above could also aid in electron and hole tunneling across the depletion region in the junction. This trap-assisted tunneling (TAT) leakage current component is very significant under reverse-bias operation (also high built-in electric field) of the diode. One can visualize TAT as electric-field-enhanced SRH current. Under this condition, electrons from the valence band of the p-type region hop into a trap in the depletion region and hop out into the conduction band of the n-type region of the diode. The tunneling probability is exponentially dependent on the physical distance of the hop. In this case, the two-step trap-assisted hop has a higher probability than a single hop from a p-type to an n-type region (as in BTBT). In advanced CMOS technologies this component of leakage current under reverse-bias operation is found to be dominant, leading to increasing leakage current with bias as opposed to a saturating behavior of the ideal diode current under reverse bias.

All three leakage phenomena discussed above are present in both forward-bias-mode and reverse-bias-mode junction currents. Under forward-bias their impact is seen only for low bias well before the diode turn on, where model accuracy is not expected. For this reason and to save on model computation time, instead of separate equations, a correction to the junction diode current equation discussed in the previous section is proposed in the form of a nonideality factor. This nonideality factor is captured through the parameter NJS used in Equation (9.3). However, under reverse-bias operation, where the leakage currents are significant (in terms of magnitude as well as bias-dependent behavior) compared with the diode reverse saturation current, a semiempirical model is used to capture both the SRH and the TAT leakage currents.

The field-enhanced SRH current or the TAT current is given by [2, 3]

$$I_{\text{jn,TAT}} = \int\limits_{\text{depletion region}} K_{\text{SRH}} \cdot (1 + \Gamma_{\text{TAT}}) \cdot \frac{\left(e^{\frac{V_{\text{jn}}}{kT}} - 1\right)}{\left(e^{\frac{\psi + 0.5V_{\text{jn}}}{kT}} + e^{\frac{-\psi + 0.5V_{\text{jn}}}{kT}} + 2\right)} \cdot d\psi, \qquad (9.13)$$

where K_{SRH} accounts for the SRH electron/hole trap capture cross-section and trap density in the depletion region, Γ_{TAT} is a field enhancement factor in order to account for the TAT leakage current, and V_{jn} is the bias voltage across the junction. Unfortunately, the integral in Equation (9.13) does not have an analytic solution, and we have to resort to some approximations to obtain one. Although it has been shown in [3] that Γ_{TAT} is a field-dependent function, we will approximate it to be a constant here. The term in the denominator inside the integral can be replaced with its maximum value that occurs at the metallurgical junction. One can then obtain a closed-form expression for the integral, and the SRH + TAT junction current is given by

$$I_{\text{jn,TAT}} = K_{\text{SRH}} \cdot (1 + \Gamma_{\text{TAT}}) \cdot W_{\text{dep}} \cdot \left(e^{\frac{V_{\text{jn}}}{2kT}} - 1\right), \qquad (9.14)$$

where W_{dep} is the depletion width. As W_{dep} is a weak function of the bias across the junction compared with the rest of the terms, not much accuracy is lost if we assume it to be a constant as well. In the BSIM-CMG model, the loss of accuracy due to the assumptions made so far in our derivation is reclaimed by introducing tuning parameters. The reverse junction leakage current for the source-side junction in the BSIM-CMG model is given by

$$I_{\mathrm{jn,es,TAT}} = -\mathrm{ASEJ} \cdot \mathrm{JTSS} \cdot \left(e^{\frac{qV_{es}}{\mathrm{NJTS} \cdot kT} \frac{\mathrm{VTSS}}{\mathrm{VTSS} - V_{es}}} - 1 \right)$$

$$- \mathrm{PSEJ} \cdot \mathrm{JTSSWS} \cdot \left(e^{\frac{qV_{es}}{\mathrm{NJTSSW} \cdot kT} \frac{\mathrm{VTSSWS}}{\mathrm{VTSSWS} - V_{es}}} - 1 \right)$$

$$- \mathrm{NF} \cdot \mathrm{NFIN} \cdot W_{\mathrm{eff0}} \cdot \mathrm{JTSSWGS} \cdot \left(e^{\frac{qV_{es}}{\mathrm{NJTSSWG} \cdot kT} \frac{\mathrm{VTSSWGS}}{\mathrm{VTSSWGS} - V_{es}}} - 1 \right), \qquad (9.15)$$

where JTSS, JTSSWS, and JTSSWGS are the pre-exponential tuning factors for the bottom area, STI side perimeter, and gate side perimeter junctions. NJTS, NJTSW, and NJTSWG are the corresponding nonideality factors, whose ideal value as observed in Equation (9.14) is equal to 2. V_{es} is the junction substrate to source terminal bias voltage. The term within the exponential is multiplied by an additional bias-dependent factor of the form $\mathrm{VTX}/(\mathrm{VTX} - V_{es})$ to ensure that the effect of the exponential terms rapidly decrease to zero under forward-bias operation. This is so because the influence of the SRH leakage component has already been taken care of through the nonideality factor NJS in Equation (9.3). Temperature dependence of the SRH + TAT leakage component has been introduced through the tuning parameters used in Equation (9.15) and will be discussed later. Drain-side reverse leakage current can be described by an analogous set of equations in a manner similar to that above.

9.2 JUNCTION DIODE CHARGE/CAPACITANCE MODEL

As described in the introduction to this chapter, the junction capacitance behavior of FinFETs has been observed to be different from that of planar bulk MOSFETs owing to the formation of a double junction in the presence of a PTS implant below the fin region. For example, Figure 9.4 shows that the reverse-bias junction capacitance $1/C_{\mathrm{jn}}^2$ versus V_{jn} curve for FinFET source-drain junctions with PTS deviates from the linear behavior of FinFET source-drain junctions without PTS (ideal step junction). The slope of this curve is inversely proportional to the *n*-type doping at the edge of the depletion region. FinFET source-drain junctions tend to show two different slopes and a higher junction capacitance when a PTS implant is used. In what follows, we will try to capture this behavior in a new junction diode charge/capacitance model.

9.2.1 REVERSE-BIAS MODEL

In a manner similar to the discussion of the junction diode current, we will restrict our discussion to a source-side junction only. The assumptions and derivations for

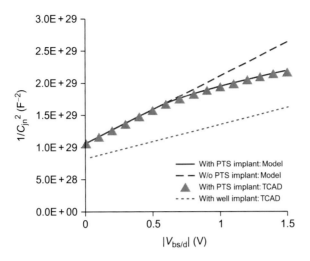

FIGURE 9.4

The $p^+|n_{pts}|n_{well}$ junction exhibits two slopes in a $1/C_{jn}^2$ versus V_{jn} plot in comparison with the single slope of the ideal step junction. The doping values used were $p^+ = 3 \times 10^{20}\,\mathrm{cm}^{-3}$, $n_{pts} = 10^{18}\,\mathrm{cm}^{-3}$, and $n_{well} = 3 \times 10^{18}\,\mathrm{cm}^{-3}$. Lines represent the results obtained using the model, and symbols represent the results obtained using TCAD.

the drain-side junction are similar. First we will derive a junction charge model for the well-known simple case of a single junction p|n diode capacitance model for reverse bias [2]. From the Poisson equation describing the electrostatics of the charge carriers in the p- and n-type regions for a diode, the junction depletion depth is given by

$$W_{dep,jn} = W_{dep0} \cdot \left(1 - \frac{V_{es}}{PBS}\right)^{MJS}, \qquad (9.16)$$

where V_{es} is the voltage across the source-side junction, W_{dep0} is the zero-bias junction depletion width, PBS represents the source-side junction built-in voltage, and MJS is the junction doping grading coefficient (which is equal to 0.5 for an ideal step junction). The junction capacitance is then given as

$$C_{jes} = \frac{\varepsilon_0 \varepsilon_r}{W_{dep,jn}} = \frac{C_{jes0}}{\left(1 - \frac{V_{es}}{PBS}\right)^{MJS}}, \qquad (9.17)$$

where ε_r is the relative dielectric constant of the junction material and C_{jes0} is the zero-bias junction capacitance. The junction depletion charge density is obtained by integrating the above equation over voltage as follows:

$$Q_{es} = \frac{C_{jes0}PBS}{1 - MJS}\left[1 - \left(1 - \frac{V_{es}}{PBS}\right)\right]^{1-MJS}. \qquad (9.18)$$

In order to capture the effect of the double junction especially the observed change in slope of the capacitance as shown before, the above charge model can be enhanced as follows:

$$
Q_{\text{es,rev}} = \begin{cases} C_{\text{j01}} \text{PBS} \dfrac{1-\left(1-\frac{V_{\text{es}}}{\text{PBS}}\right)^{1-\text{MJS}}}{1-\text{MJS}} & 0 < V_{\text{es}} < V_{\text{ec}}, \\[2em] C_{\text{j01}} \text{PBS} \dfrac{1-\left(1-\frac{V_{\text{es}}}{\text{PBS}}\right)^{1-\text{MJS}}}{1-\text{MJS}} + C_{\text{j02}} \text{PBS2} \dfrac{1-\left(1-\frac{V_{\text{es}}-V_{\text{ec}}}{\Phi_{\text{b2}}}\right)^{1-\text{MJS2}}}{1-\text{MJS2}} & V_{\text{ec}} < V_{\text{es}}, \end{cases}
$$

$$(9.19)$$

where C_{j01} and C_{j02} are the zero-bias capacitance values and PBS and PBS2 are the barrier height of the $p^+|n_{\text{well}}$ and $p^+|n_{\text{pts}}$ junctions. MJS and MJS2 represent the gradient of the $p^+|n_{\text{pts}}$ and $n_{\text{pts}}|n_{\text{well}}$ junctions. We can observe that the first term in Equation (9.19) is the same as that for a single junction diode. Equation (9.19) maintains charge continuity at the crossover voltage $V_{\text{es}} = V_{\text{ec}}$. The continuity of the first and second derivatives of charge also needs to be ascertained for accuracy in the prediction of higher harmonic power content in the output of a transistor in analog/RF circuit simulations. The continuity of the first derivative of charge in Equation (9.19) (which is also the junction capacitance) at $V_{\text{es}} = V_{\text{ec}}$ yields

$$
C_{\text{j01}} \left(1 - \frac{V_{\text{ec}}}{\text{PBS}}\right)^{-\text{MJS}} = C_{\text{j02}}.
$$

$$(9.20)$$

Ensuring continuity of the second derivative of the charge (first derivative of capacitance) at $V_{\text{es}} = V_{\text{ec}}$ gives rise to the following condition:

$$
C_{\text{j01}} \text{MJS} \frac{\left(1 - \frac{V_{\text{ec}}}{\text{PBS}}\right)^{-1-\text{MJS}}}{\text{PBS}} = \frac{C_{\text{j02}} \text{MJS2}}{\text{PBS2}}.
$$

$$(9.21)$$

These conditions, Equations (9.20) and (9.21), are factored into the parameter extraction process. In the junction capacitance curve, $1/C_{\text{jn}}^2 - V_{\text{es}}$ (Figure 9.4), the first slope region corresponding to the depletion edge traversing the PTS implant region is used to extract the values for the parameters C_{j01}, PBS, and MJS in a similar way as for a single junction diode. Among the four remaining parameters (V_{ec}, C_{j02}, PBS2, and MJS2) that correspond to the n_{well} region (the second slope region), Equations (9.20) and (9.21) allow us the flexibility to choose only two of them. We chose C_{j02} and PBS2, which signify the depth of the PTS implant n_{well} region boundary and the n_{well} region doping concentration. The parameters V_{ec} and PBS2 are then determined by simultaneously solving Equations (9.20) and (9.21) using the values chosen for C_{j02} and MJS2:

$$
V_{\text{ec}} = \text{PBS} \cdot \left[1 - \left(\frac{C_{\text{j01}}}{C_{\text{j02}}}\right)^{\frac{1}{\text{MJS}}}\right],
$$

$$(9.22)$$

$$
\text{PBS2} = \frac{\text{PBS} \cdot \text{MJS2} \cdot C_{\text{j02}}}{C_{\text{j01}} \cdot \text{MJS} \cdot \left(1 - \frac{V_{\text{ec}}}{\text{PBS}}\right)^{-1-\text{MJS}}}.
$$

$$(9.23)$$

The junction capacitance is then given by the derivative dQ_{es}/dV_{es} as follows:

$$C_{jes,rev} = \begin{cases} C_{j01} \left(1 - \frac{V_{es}}{PBS}\right)^{-MJS} & 0 < V_{es} < V_{ec}, \\ C_{j02} \left(1 - \frac{V_{es} - V_{ec}}{PBS2}\right)^{-MJS2} & V_{ec} < V_{es}. \end{cases} \tag{9.24}$$

To validate the model, the structure in Figure 9.2 was simulated using TCAD [4]. A source/drain region p^+ doping was chosen as $3 \times 10^{20}\,\mathrm{cm}^{-3}$. The dopings for the n_{pts} and n_{well} regions were chosen to be $10^{18}\,\mathrm{cm}^{-3}$ and $3 \times 10^{18}\,\mathrm{cm}^{-3}$, respectively. The two terminal capacitances across the source/drain and substrate were extracted. From Figure 9.4 we can observe that the crossover voltage, V_{bc}, is around 0.6 V of reverse-bias junction voltage, where the slope of the $1/C_{jn}^2$ plot changes, reflecting the doping at the edge of the depletion region. The derived model in Equation (9.24) shows excellent agreement (after parameter tuning) with TCAD simulations for such a junction (Figure 9.4).

9.2.2 FORWARD-BIAS MODEL

The charge/capacitance model in Equations (9.19) and (9.24) is not valid for voltages V_{es} approaching—PBS—i.e., when the diode is forward biased. In a realistic diode, series resistances of the quasi-neutral regions dominate and restrict the actual voltage drop across the junction (or rather across the depletion edges on either side of the metallurgical junction). Equation (9.24) would result in very large values, making it unwieldy for implementation in a compact modeling framework. For this reason, industry standard compact models resort to an approximation for the forward-bias region. For example, BSIM4 resorts to using a quadratic equation to describe the junction charge for the forward bias (i.e., junction capacitance is made linear). For a p-type FET, the forward-bias junction charge is essentially given by a Taylor series expansion of Equation (9.24) up to second order around $V_{es} = 0\,\mathrm{V}$:

$$Q_{es,fwd} = C_{j01} \cdot V_{es} + \frac{C_{j01} MJS}{2 \cdot PBS} V_{es}^2 \quad \text{for } V_{es} < 0. \tag{9.25}$$

Equation (9.25) together with Equation (9.19) ensures the continuity of junction charge and its first derivative capacitance around $V_{es} = 0\,\mathrm{V}$. However, there is a discontinuity in the third derivative of the overall junction charge, i.e., $d^3 Q_{es}/dV_{es}^3$ at $V_{bs/d} = 0\,\mathrm{V}$. This discontinuity is not a cause for concern because of convergence of a compact model. SPICE simulators require continuity of up to only the second derivative of charge for convergence. However, for the RF design community that is interested in accurate prediction of higher-order intermodulation distortion products to accurately predict out-of-band emission by wireless transmitters, continuity of charge up to sixth order is desired. For example, in the operation of a passive mixer, the FET operates in a strict linear region with source-drain voltage $V_{ds} = 0\,\mathrm{V}$. For this mixer, if the FET body is tied to ground, the voltage across the source-drain junction is also centered around 0V during the mixing operation. The discontinuity when Equations (9.19) and (9.25) are put together will lead to incorrect predictions of

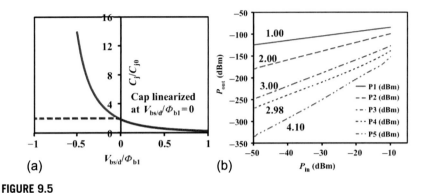

FIGURE 9.5

(a) The first derivative of the junction capacitance using the linearization in Equation (9.25). (b) Results of the harmonic balance test for a *p*-type MOS passive mixer configuration showing wrong slopes for the fourth- and fifth-harmonic output power.

higher-order intermodulation products. This can be verified using a simple harmonic-balance simulation setup wherein a single-tone RF stimulation is supplied to the source of a FET with its gate voltage set to above the threshold voltage. The power in the harmonic content of the drain current is observed. The slope of the nth-harmonic component power versus input power will be n at low power input. Any discontinuity or issues with model symmetry with respect to the source and the drain would lead to wrong slopes. In Figure 9.5, we report results from such a test for a model with a symmetric core but the above junction charge/capacitance model. As expected, we see that deviation in the slope for the fourth- and fifth-harmonic output power. In order to rectify this, we propose an alternative model as follows. Instead of a Taylor series expansion around $V_{es} = 0$ V, pushing the transition point further into forward bias helps. We choose a quadratic Taylor series expansion around $V_{es} = k \cdot PBS$ for this as follows:

$$Q_{es,fwd} = C_{j01} PBS \frac{(1-k)^{1-MJS}}{1-MJS} + C_{j01}(1-k)^{-MJS} \cdot (V_{es} + k \cdot PBS) \cdots$$

$$+ C_{j01} MJS \frac{(1-k)^{-1-MJS}}{2 \cdot PBS} \cdot (V_{es} + k \cdot PBS)^2 \quad \text{for } V_{es} < -k \cdot PBS. \quad (9.26)$$

For implementation purposes in BSIM-CMG we have chosen $k = 0.9$. This change results in improved accuracy to higher-order intermodulation products. This modified model was implemented for junction capacitance in BSIM-CMG, and results from the harmonic-balance simulations for a passive mixer are shown in Figure 9.6. As expected, we see that this model predicts accurate slopes up to fifth-harmonic content in the MOSFET. The persisting discontinuity at very high forward bias should not be a problem. Given that the diode current behaves exponentially with bias voltage V_{es}, all the voltage drop will be across the parasitic source-drain resistance or the substrate network resistance, restricting the diode junction from seeing such high voltages.

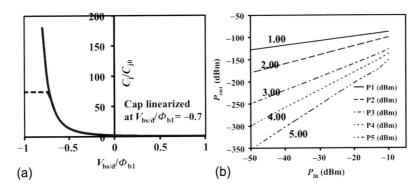

FIGURE 9.6

(a) The first derivative of junction capacitance using the linearization in Equation (9.26) with $k = 0.7$. (b) Results of the harmonic-balance test for a p-type MOS passive mixer configuration showing correct slopes for the fourth- and fifth-harmonic output power.

The overall junction charge model is as follows:

$$
Q_{es} = \begin{cases}
C_{j01} \text{PBS} \dfrac{(1-k)^{1-\text{MJS}}}{1-\text{MJS}} + C_{j01}(1-k)^{\text{MJS}} \cdot (V_{es} + k \cdot \text{PBS}) \cdots \\[2mm]
+ C_{j01} \text{MJS} \dfrac{(1-k)^{-1-\text{MJS}}}{2 \cdot \text{PBS}} \cdot (V_{es} + k \cdot \text{PBS})^2 & V_{es} < -k \cdot \text{PBS}, \\[4mm]
C_{j01} \text{PBS} \dfrac{1 - \left(1 - \frac{V_{es}}{\text{PBS}}\right)^{1-\text{MJS}}}{1 - MJS} & -k \cdot \text{PBS} < V_{es} < V_{ec}, \\[4mm]
C_{j01} \text{PBS} \dfrac{1 - \left(1 - \frac{V_{ec}}{\text{PBS}}\right)^{1-\text{MJS}}}{1 - \text{MJS}} + C_{j02} \text{PBS2} \dfrac{1 - \left(1 - \frac{V_{es} - V_{ec}}{\text{PBS2}}\right)^{1-\text{MJS2}}}{1 - \text{MJS2}} & V_{ec} < V_{es}.
\end{cases}
$$

(9.27)

The junction capacitance given by dQ_{es}/dV_{es} is given by

$$
C_{jes} = \begin{cases}
\dfrac{C_{j01}}{(1-k)^{\text{MJS}}} + \dfrac{C_{j01} \text{MJS}}{\text{PBS}(1-k)^{1+\text{MJS}}} \cdot (V_{es} + k \cdot \text{PBS}) & V_{es} < -k \cdot \text{PBS}, \\[3mm]
C_{j01}\left(1 - \dfrac{V_{es}}{\text{PBS}}\right)^{-\text{MJS}} & -k \cdot \text{PBS} < V_{es} < V_{ec}, \\[3mm]
C_{j02}\left(1 - \dfrac{V_{es} - V_{ec}}{\text{PBS2}}\right)^{-\text{MJS2}} & V_{ec} < V_{es}.
\end{cases}
$$

(9.28)

Similarly to the junction diode current model, junction diode capacitance is split into three components as well—bottom area, STI perimeter, and gate-edge components. The derivation so far was a generic description for each of the three components. Separate sets of parameters are used to describe the charge for each of the components. C_{j01} in the above equations is replaced with ASEJ $\cdot$ CJS, PSEJ $\cdot$ CJSWS, and NF $\cdot$ NFIN $\cdot$ W_{eff0} $\cdot$ CJSWGS for the bottom area, STI perimeter, and gate-edge components, respectively.

Similarly PBS, PBSWS, and PBSWGS represent the built-in potentials and MJS, MJSWS, and MJSWGS represent the junction doping gradient for the corresponding components, respectively. In the BSIM-CMG model, C_{j02} is described relative to C_{j01} using a new parameter: $C_{j02} = \text{SJS} \cdot C_{j01}$. SJS, SJSWS, and SJSWGS would be the parameters that correspond to the three diode components for this purpose. MJS2, MJSWS2, and MJSWGS2 are the parameters describing the doping gradient of the second junction for all the components. If in a technology the effect of the second junction is not very strong, one can conveniently set SJS, SJSWS, and SJSWGS to zero. This would convert the model to the default single step junction diode capacitance.

REFERENCES

[1] K. Okano, T. Izumida, H. Kawasaki, A. Kaneko, A. Yagishita, T. Kanemura, M. Kondo, S. Ito, N. Aoki, K. Miyano, T. Ono, K. Yahashi, K. Iwade, T. Kubota, T. Matsushitaand, I. Mizushima, S. Inaba, K. Ishimaru, K. Suguro, K. Eguchi, Y. Tsunashima, H. Ishiuchi, Process integration technology and device characteristics of CMOS FinFET on bulk silicon substrate with sub-10 nm fin width and 20 nm gate length, in: International Electron Devices Meeting Technical Digest (IEDM), 2005, pp. 721–724.
[2] S.M. Sze, K.K. Ng, Physics of Semiconductor Devices, Wiley, New York, 2006.
[3] G.A.M. Hurkx, D.B.M. Klaassen, M.P.G. Knuvers, A new recombination model for device simulation including tunneling, IEEE Trans. Electron Devices 39 (2) (1992) 331–338.
[4] Sentaurus Device, Synopsys, Inc., 2013.

Benchmark tests for compact models

CHAPTER OUTLINE

10.1 Asymptotic correctness... 218
10.2 Benchmark tests.. 219
    10.2.1 Tests for checking physical behavior in weak-inversion and
            strong-inversion regions.. 219
    10.2.2 Symmetry tests 222
    10.2.3 Reciprocity test for capacitances in a compact model.............. 226
    10.2.4 Test for the self-heating effect model.............................. 227
    10.2.5 Tests for the thermal noise model 227
References... 228

As discussed in Chapter 1, compact models are critically important for circuit simulations. They must not only meet the accuracy requirement, but must also be mathematically and physically robust for the convergence of the simulations. Circuit simulators, while trying to find the solutions of the Kirchhoff current laws and Kirchhoff voltage laws of a network, continuously use the compact models of the components present in the network. A well-behaved compact model increases the chance of finding the solution. It is thus very important to test a compact model for its robustness with respect to convergence before deploying it in a circuit simulator. Testing a compact model becomes particularly important if it is to be used in RF and analog circuit design. This is because RF/analog circuit simulations require a more advanced type of simulation algorithm, such as harmonic balance, and these force more stringent requirements on the compact models for convergence. So, how can one test the compact model for convergence? One simple check of the quality of compact models is via the principle of asymptotic correctness. This checks if the model behaves as expected in extreme input conditions. Another, more involved way to evaluate the physical behavior of a compact model is via several benchmark tests developed over the years [1–7]. These tests have been designed to check different aspects of a compact model. We discuss both the asymptotic correctness principle and different benchmark tests in this chapter. The BSIM-CMG model is taken as an example, and all the benchmark tests are performed on it.

FinFET Modeling for IC Simulation and Design. http://dx.doi.org/10.1016/B978-0-12-420031-9.00010-5

10.1 ASYMPTOTIC CORRECTNESS

This is a simple principle to check for potential problems in a compact model. As the model inputs such as device geometry, temperature, or bias go to extreme values (both extremely large and extremely small), the model equations should behave in a physical manner and should not have any numerical problems. The following simple examples illustrate this principle.

The forward bias diode current increases exponentially with the applied voltage, and can be represented by the expression

$$I = I_0 e^{V_a/V_{th}}. \tag{10.1}$$

Here V_a is the applied voltage, V_{th} is the thermal voltage, and I_0 is a model parameter. Although this equation can fit the measured data very well, it will predict nonzero current at zero applied bias! This is clearly unphysical and a clear violation of the principle of asymptotic correctness. A better model for the diode current is

$$I = I_0 \left(e^{V_a/V_{th}} - 1 \right). \tag{10.2}$$

Equation (10.2) ensures there is zero current for zero bias and is an asymptotically correct model.

Another example is the modeling of the device geometry dependence of the thermal resistance (R_{TH}) in silicon-on-insulator MOSFETs. R_{TH} decreases with an increase in the area and perimeter of the device [8]. A simple way to model this would be to write

$$R_{TH} = \frac{R_{THA}}{WL} + \frac{R_{THP}}{W + L}, \tag{10.3}$$

where R_{THA} and R_{THP} are fitting parameters, and W and L are the device width and length, respectively. The denominators in Equation (10.3) represent the area (WL) and the perimeter ($W + L$) of the device. Although Equation (10.3) can fit the data for a limited set of devices, this equation will behave in an unphysical way as one of L or W decreases to an extremely small value as it would predict a very large value for the thermal resistance. Physically, R_{TH} should be finite if W or L becomes small as heat conduction can take place from the perimeter of the device. Thus, this model is asymptotically incorrect. A better approach is to think in terms of thermal conductance as

$$R_{TH} = \frac{1}{G_{TH}} = \frac{1}{G_{THA} WL + G_{THP} (W + L)}, \tag{10.4}$$

where G_{THA} and G_{THP} are fitting parameters. Equation (10.4) is well behaved asymptotically and is a better model than Equation (10.2). In summary, the principle of asymptotic correctness is about checking the physical sanity of a model under extreme input conditions of applied bias, geometry, etc.

10.2 BENCHMARK TESTS

A benchmark test is a specific type of simulation performed with the model. The output is compared with physically correct results to see how a compact model fares. Over the years, several benchmark tests have been designed to check various aspects of a compact model as discussed in following subsections.

10.2.1 TESTS FOR CHECKING PHYSICAL BEHAVIOR IN WEAK-INVERSION AND STRONG-INVERSION REGIONS

These tests check if the model behavior is physically correct in weak-inversion and strong-inversion regions. The strong-inversion region is obviously important; however, the weak-inversion region of operation is becoming increasingly important with the decreasing supply voltages in the advanced technology nodes. Also, analog and RF circuits/devices are routinely biased in the weak-inversion region (see Chapter 2), further underlining its importance. The benchmark tests for weak-inversion and strong-inversion regions are the slope ratio (SR) test, the output conductance test, and the volume inversion test. These are described below.

Slope ratio test

This test checks if a compact model takes care of the different drain-to-source voltage V_{ds} dependence of drain-to-source current I_{ds} in weak-inversion and strong-inversion regions. In the weak-inversion region, the V_{ds} dependence of I_{ds} can be expressed as

$$I_{ds} \alpha \left(1 - e^{-(V_{ds}/V_{th})}\right),\tag{10.5}$$

where V_{th} is the thermal voltage. In the strong-inversion region (linear condition) I_{ds} becomes a linear function of V_{ds}. To see if the compact model captures this important difference, SR is calculated from the weak-inversion region to the strong-inversion region. SR is defined as

$$SR = \frac{(I_{ds2} + I_{ds1})(V_{ds2} - V_{ds1})}{(I_{ds2} - I_{ds1})(V_{ds2} + V_{ds1})},\tag{10.6}$$

where I_{ds1} and I_{ds2} are drain-to-source currents at drain-to-source voltages V_{ds1} and V_{ds2}, respectively. At room temperature, SR should smoothly and monotonically decrease from a value close to 1.3 to 1.0 on moving from weak to strong inversion. This variation in SR from weak to strong inversion can be worked out by inserting the physically correct V_{ds} dependence of I_{ds} in respective regions.

In Figure 10.1, we show that the BSIM-CMG model passes this test and SR calculated from the model indeed varies smoothly between the values discussed above. Note that the value of SR in weak inversion is temperature dependent owing to the presence of thermal voltage V_{th} in Equation (10.5).

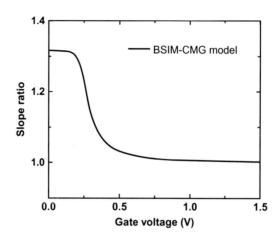

FIGURE 10.1

SR calculated using the BSIM-CMG model at room temperature. $V_{ds1} = 10$ mV and $V_{ds2} = 20$ mV were used for the calculations. $T_{fin} = 15$ nm, $L = 1$ μm, and $N_{fin} = 10$.

Conductance test

This test also checks if the model captures the difference in behavior of I_{ds} with V_{ds} in weak-inversion and strong-inversion regions. From Equation (10.5) it can be seen that in the weak-inversion region

$$\frac{\partial}{\partial V_{ds}} \left[g_{ds} \exp \left(V_{ds}/V_{th} \right) \right] = 0, \qquad (10.7)$$

where $g_{ds} = \partial I_{ds}/\partial V_{ds}$ is the output conductance. Hence, if $g_{ds} \exp(V_{ds}/V_{th})$ is plotted versus V_{ds} in the weak-inversion region, the line should be absolutely flat. As we move from weak to strong inversion, there is a finite slope seen in the plot of $g_{ds} \exp(V_{ds}/V_{th})$ versus V_{ds}. In Figure 10.2, we show that the BSIM-CMG model behaves in the physically correct manner.

Volume inversion test

In the subthreshold region of thin-body lightly doped multigate devices, the gate bias moves the energy bands not just at the surface but across the full body thickness of the device [9, 10]. This causes inversion to occur in the full body thickness of the device as opposed to only at the surface in the case of bulk MOSFETs. This phenomenon is called volume inversion. It occurs only in the subthreshold region as in strong inversion the carriers at the surface screen the electric field from reaching deep inside the silicon body. As a result of volume inversion, the subthreshold current in these devices increases with an increase in the fin thickness T_{fin}. This phenomenon should be captured by the compact model. A compact model can be tested to find if it does capture volume inversion by obtaining the subthreshold current of a long-channel

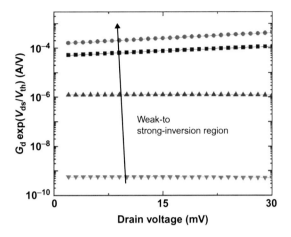

FIGURE 10.2

Result of the output conductance test in weak- and strong-inversion regions using the BSIM-CMG model. $V_g = 0$–0.6 V in steps of 0.2 V. $T_{fin} = 15$ nm, $L = 1$ μm, and $N_{fin} = 10$.

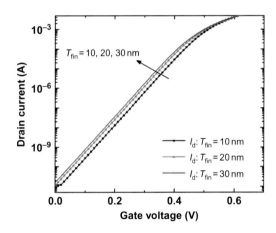

FIGURE 10.3

Subthreshold current obtained from the BSIM-CMG model. The current increases with an increase in fin thickness T_{fin} owing to the volume inversion effect.

device for various fin thicknesses. A long-channel device is used to avoid any complication with results coming from short-channel effects. All the real device effects (see Chapter 4) should be turned off in the model. The subthreshold current should increase with increasing fin thickness. In Figure 10.3, we show simulations from the BSIM-CMG model for device fin thicknesses $T_{fin} = 10$, 20, and 30 nm. The subthreshold current increases with the increase in T_{fin}, demonstrating that the BSIM-CMG model captures the volume inversion phenomenon.

10.2.2 SYMMETRY TESTS

Physically, a MOSFET is symmetric with respect to the source and drain terminals. Ideally this characteristic should be reflected by the compact model. However, there can be approximations in the compact model which do not treat the drain and the source in the same way. For example, the body effect is often modeled as a function of body-to-source voltage, but not body-to-drain voltage. This may cause a violation of source-drain symmetry by the compact model. It has been shown that simulations of circuits such as pass gates are not even qualitatively correct if a compact model violates this source-drain symmetry requirement [11]. The benchmark tests which can be used to check the model symmetry are described in this section along with the results for the BSIM-CMG model for each of these tests.

Gummel symmetry test

The Gummel symmetry test is a standard way of testing the symmetry of the drain-current model. The simulation setup for this test is shown in Figure 10.4. A voltage source V_x is applied at the drain terminal and $-V_x$ is applied at the source terminal. V_x is swept from a negative value to a positive value (typically 0.1 V) for a set of V_g and V_b. If the model is symmetric, it should satisfy

$$I_x\left(V_g, V_x, -V_x, V_b\right) = -I_x\left(V_g, -V_x, V_x, V_b\right) \qquad (10.8)$$

It follows from Equation (10.8) that the odd-order derivatives of I_x with respect to V_x are even functions, while the even-order derivatives are odd functions of V_x. In addition to checking the symmetry in I_x, this test is also used to see the continuity and smoothness of I_x. This can be checked by plotting first-, second-, third-, and higher-order derivatives of I_x with respect to V_x obtained from this setup. A well-behaved compact model should have smooth and continuous derivatives of I_x at least up to fifth order. Smoothness and continuity of derivatives is especially important

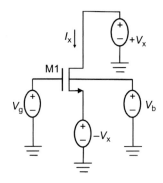

FIGURE 10.4

Setup for the Gummel symmetry test. M1 is the device under test simulated using its compact model.

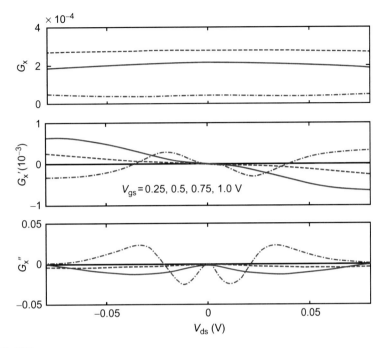

FIGURE 10.5

Results of the Gummel symmetry test performed on the BSIM-CMG model. The model
shows smooth and continuous derivatives of the drain current and passes the
symmetry test.

when the model is used in distortion analysis of circuits. In Figure 10.5, the Gummel
symmetry test results for the BSIM-CMG model are shown. The BSIM-CMG model
clearly passes the Gummel symmetry test.

Harmonic balance simulation test

This test is very important for RF circuits as harmonic balance simulations are
routinely used in RF circuit design. It is well known that if a compact model
has singularities in drain-current derivatives it will predict unphysical harmonics.
Theoretically, the second harmonic is proportional to the square of the input signal,
the third harmonic is proportional to the cube of the input signal, and so on [12].
A well-behaved compact model should mimic this behavior when harmonic balance
simulations are performed. To test a compact model for this, the setup shown in
Figure 10.6 can be used. An RF signal is applied to the source/drain of the device
and the harmonic components of the drain current are plotted as a function of the
input RF signal. In Figure 10.7, harmonic balance simulation results from the BSIM-
CMG model are shown to be in good agreement with the theory.

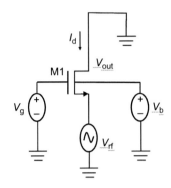

FIGURE 10.6

Setup for harmonic balance simulations.

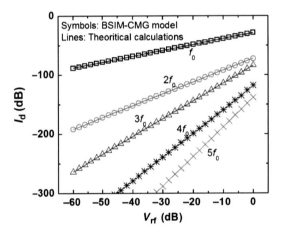

FIGURE 10.7

Harmonic balance simulation results showing the fundamental frequency f_0, second harmonic $2f_0$, third harmonic $3f_0$, fourth harmonic $4f_0$, and fifth harmonic $5f_0$ obtained from the BSIM-CMG model. The model predictions agree very well with the theoretical calculations. $f_0 = 1$ MHz and $V_g = 1$ V.

AC symmetry test

In addition to the drain current, the terminal charges obtained from the compact model should also be symmetric with respect to the source and drain terminals. The setup shown in Figure 10.8 can be used to check the symmetry of the charge models. AC sources in phase and out of phase are applied to the source and the drain of the device, and terminal currents are monitored as shown in Figure 10.8. DC voltages V_x and $-V_x$ are applied to the drain and the source terminals of the device, respectively. The

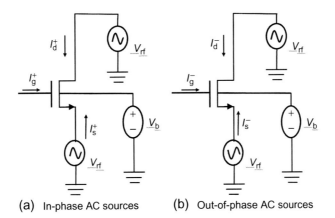

(a) In-phase AC sources   (b) Out-of-phase AC sources

FIGURE 10.8

AC symmetry test setup showing AC currents for AC sources that are in phase (a) and out of phase (b) with each other. DC bias V_x is applied at the drain and $-V_x$ is applied at the source terminal of the device.

imaginary part of the AC terminal current i_g for in-phase (i_g^+) and out-of-phase (i_g^-) sources is then given by

$$i_g^+ = 2\pi f \left(C_{gs} + C_{gd}\right),\tag{10.9a}$$

$$i_g^- = 2\pi f \left(C_{gs} - C_{gd}\right),\tag{10.9b}$$

respectively, where f is the frequency of the applied AC signal, and C_{gs} and C_{gd} are gate-to-source and gate-to-drain capacitances, respectively. Then the quantity δC_g defined as

$$\delta C_g = \frac{i_g^-}{i_g^+} = \frac{C_{gs} - C_{gd}}{C_{gs} + C_{gd}}\tag{10.10}$$

should be an odd function of V_x. δC_g checks for the gate-charge model. In a similar way, drain-charge and source-charge models can be tested by defining δC_{sd} given by

$$\delta C_{sd} = \frac{C_{ss} - C_{dd}}{C_{ss} + C_{dd}},\tag{10.11}$$

where C_{ss} and C_{dd} are source and drain capacitances, respectively.

In Figure 10.9 we show the results of the BSIM-CMG model for the AC symmetry tests described above. We plot δC_g, δC_{sd}, and their first derivatives, which show symmetry of the charge model as well as its continuous and smooth behavior.

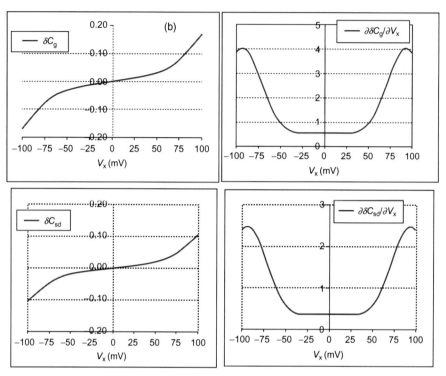

FIGURE 10.9

δC_g and δC_{sd} versus V_x calculated from the BSIM-CMG model demonstrating the symmetry of the charge model. Derivatives of δC_g and δC_{sd} show that the charge model in the BSIM-CMG model is continuous and smooth.

10.2.3 RECIPROCITY TEST FOR CAPACITANCES IN A COMPACT MODEL

Surface-potential or charge-based compact models describe the device capacitance behavior by developing analytical equations of the terminal charges of the device as shown in Chapter 6. The terminal charge Q_i (of the ith terminal) can then be used to evaluate the capacitance between any two terminals i and j of the device by the following expression:

$$C_{ij} = \delta_{ij} \frac{\partial Q_i}{\partial V_j}, \quad \text{where } \delta_{ij} = 1 \text{ when } i = j, \text{ else } \delta_{ij} = -1. \quad (10.12)$$

The capacitances obtained from the compact model should behave in a physical way. One such requirement is the so-called reciprocity of capacitances at $V_{ds} = 0$. Reciprocity means the model should satisfy $C_{ij} = C_{ji}$ at $V_{ds} = 0$. This is shown to be the case for C_{gs} and C_{sg} obtained from the BSIM-CMG model in Figure 10.10. The same behavior is also observed for C_{gd} and C_{dg}.

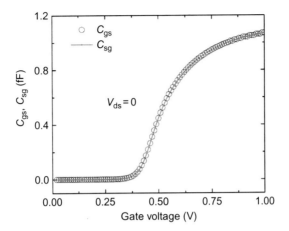

FIGURE 10.10

Capacitance (C_{gs}, C_{sg}) versus V_g obtained from the BSIM-CMG model. $C_{gs} = C_{sg}$ at $V_{ds} = 0$ demonstrates that the BSIM-CMG model has reciprocal capacitances at $V_{ds} = 0$. Other pairs of capacitances also pass the reciprocity test.

10.2.4 TEST FOR THE SELF-HEATING EFFECT MODEL

The self-heating effect is modeled via a thermal network consisting of thermal resistance and thermal capacitance [13]. The voltage across the thermal network is numerically equal to the local channel temperature T_{ch} of the device. To test the implementation of the self-heating model, the following simulations should be performed. First, with the self-heating effect turned on, sweep the thermal resistance in the model and obtain the values of the drain current I_d and channel temperature T_{ch} at fixed V_{gs} and V_{ds}. Next, self-heating should be disabled, and for the same V_{gs} and V_{ds} condition as in the previous simulation, I_d should be obtained by sweeping the ambient temperature T_{amb}. For $T_{ch} = T_{amb}$, the drain current obtained in the two scenarios should be exactly the same. In Figure 10.11, we show the results of this test from the BSIM-CMG model.

10.2.5 TESTS FOR THE THERMAL NOISE MODEL

A complete compact model also includes a model for thermal noise behavior of the device. A simple test to check the thermal noise model is as follows. Bias the device at $V_{ds} = 0$ and a gate voltage well above the threshold voltage of the device. For such a bias condition, the device is essentially a resistor of value $R = 1/g_{ds}$. Run the noise simulations at low enough frequencies where capacitances are not important. The current noise spectral density obtained from the model should be equal to $4kTg_{ds}$ A²/Hz. The BSIM-CMG model for $g_{ds} = 10^{-4}$ A/V produces 1.65×10^{-24} A²/Hz as the current noise spectral density at the drain node, which matches the theoretical calculations.

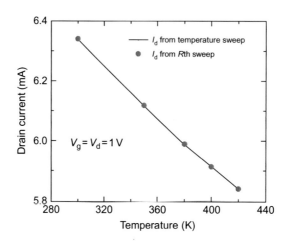

FIGURE 10.11

Self-heating test results for the BSIM-CMG model. $T_{fin} = 15$ nm, $L = 30$ nm, $N_{fin} = 100$, and $V_g = V_d = 1$ V. The values of I_d obtained in the two different simulation setups match very well. This shows that the BSIM-CMG model passes the self-heating test.

REFERENCES

[1] Y. Tsividis, Operation and Modeling of the MOS Transistor, second ed., Oxford University Press, Oxford, 1999.

[2] C. McAndrew, Practical modeling for circuit simulations, IEEE J. Solid State Circuits 33 (3) (1998) 439–448.

[3] Y. Tsividis, K. Suyama, MOSFET modeling for analog circuit CAD: problems and prospects, IEEE J. Solid State Circuits 29 (3) (1994) 210–216.

[4] C. McAndrew, H. Gummel, K. Singhal, Benchmarks for compact MOSFET models, Proceedings. SEMATECH Compact Models Workshop, 1995.

[5] X. Li, W. Wu, A. Jha, G. Gildenblat, R. van Langevelde, G.D.J. Smit, A.J. Scholten, D.B.M. Klassen, C.C. McAndrew, J. Watts, M. Olsen, G. Coram, S. Chaudhary, J. Victory, "Benchmarking the PSP compact model for MOS transistors", Proceedings of the IEEE ICMTS, 2007, pp. 259–264.

[6] C. McAndrew, "Validation of MOSFET model source drain symmetry", IEEE Trans. Electron Dev. 53 (9) (2006) 2202–2206.

[7] G. Gildenblat, Compact Modeling Principles, Techniques and Applications, Springer, Berlin, 2010.

[8] S. Khandelwal, J. Watts, E. Tamilmani, L. Wagner, Scalable thermal resistance model for single- and multi-finger SOI MOSFETs, Proceedings IEEE ICMTS, 2011, pp. 182–185.

[9] Y. Taur, An analytical solution to a double gate MOSFET with undoped body, IEEE Electron Dev. Lett. 21 (5) (2005) 245–247.

[10] Y. Taur, X. Liang, W. Wang, H. Lu, A continuous, analytic drain current model for DG MOSFETs, IEEE Electron Dev. Lett. 25 (2) (2004) 107–109.

[11] K. Joardar, K.K. Gullapalli, C.C. McAndrew, M.E. Burnham, A. Wild, An improved MOSFET model for circuit simulation, IEEE Trans. Electron Dev. 45 (1) (1998) 104–148.

[12] T.H. Lee, The Design of CMOS Radio Frequency Integrated Circuits, second ed., Cambridge University Press, Cambridge, 2004.

[13] BSIMSOI4.6 Users' Manual, BSIM Group. Available: http://www-device.eecs.berkeley.edu/bsim/.

BSIM-CMG model parameter extraction

CHAPTER OUTLINE

11.1 Parameter extraction background 232
11.2 BSIM-CMG parameter extraction strategy 233
11.3 Conclusion 242
References 242

Compact models contain many parameters to model various effects in the device as described in Chapter 4. Model equations are parameterized for the following reasons:

(1) The core model derivation is for an "ideal" device; for instance, the channel doping profile is known and perfectly uniform and the mobility is known and its dependence on gate voltage is negligible. As a result, the core model cannot predict an actual transistor's characteristics, even for a long-channel transistor, accurately enough for IC design without some fitting parameters. Parameters are required to match the model to the measured characteristics of the actual device.

(2) There are many real-device effects which are practically impossible to model with sufficient accuracy and computational speed without using fitting parameters. In such cases, simple equations based on good understanding of the device physics can do wonders to capture the complex effects of voltage biases and device geometry on the device behaviors in general, and very accurately with the help of a few fitting parameters. For instance, the output resistance as a function of V_{ds}, V_{gs}, V_{bs}, and L (see Figure 11.8) could not be accurately modeled by pure "curve fitting" until physics-based model equations were introduced (see Section 4.11 of Chapter 4). One should not equate the use of fitting parameters with pure curve fitting.

(3) Model parameters are also needed because the device geometry—for example, the shape of the corner—or doping profile is not known exactly or has manufacturing variations.

In fact, judicial introduction of parameters in the model is extremely important for its practical use. Extracting the values of these parameters is thus a critical step before using the model for circuit design. Indeed, model accuracy and sometimes even its convergence properties heavily depend on the extracted values of the parameters. This chapter discusses the parameter extraction procedure for the BSIM-CMG model. Model results for measured data with channel length ranging from 30 nm to 10 μm are shown for the BSIM-CMG model.

11.1 PARAMETER EXTRACTION BACKGROUND

As discussed earlier, a compact model contains many parameters in order to capture various physical effects in the device. The parameter values are found by fitting a variety of measured data, such as $I_d - V_g$, $I_d - V_d$, $C_{gg} - V_g$, and $I_d - V_g$, at multiple ambient temperatures, for example. Extraction of parameter values is thus a multidimensional optimization problem. In fact, optimization algorithms such as the Levenberg-Marquardt algorithm [1] and the genetic algorithm [2] and particle swarm optimization are quite regularly used for parameter extraction [3, 4]. There are many commercial tools available for parameter extraction, such as BSIMProPlus [5], ICCAP [6], MBP [7], and UTMOST [8]. However, in order to use these sophisticated algorithms and tools effectively, an understanding of the model equations along with the parameters and the device characteristics they affect is very important.

Before starting a parameter extraction process, one must understand an important classification of parameters. Generally, parameters in a compact model are classified as global and local parameters. This classification can be understood as follows. Consider the case when the parameter extraction is to be performed for a single device. The parameter extraction engineer needs only tune the parameters specific to particular effects such as U0 and RDSW without exercising the geometry dependence of these parameters. The parameter set obtained after such an extraction is a "local" set as it applies to a specific geometry device. On the other hand, when parameter extraction is performed for a range of channel lengths and widths, the geometry dependence of the "local" parameters needs to be invoked. For instance, the mobility parameter U0 is found to decrease with channel length and this L dependence is modeled via AU0 and BU0. Hence, while U0 being "local" is specific to a device, the parameters AU0 and BU0 can be termed global parameters. The various length-scaling equations introduced in the BSIM-CMG model are described in [9]. In a standard CMOS technology, various device geometries for the same flavor of device—for example, n-channel high-performance (low threshold voltage) MOSFET—are manufactured to cater to different circuit design needs. A compact model should be able to predict the electrical behavior accurately for various device geometries. This is possible only when the global parameter set is extracted for the device flavor. In the next section, we discuss the global parameter extraction strategy for the BSIM-CMG model.

11.2 BSIM-CMG PARAMETER EXTRACTION STRATEGY

A complete global parameter extraction flow is shown in Figure 11.1. It starts from parameter initialization to the extraction of noise model parameters. Here, we discuss mainly the drain-current parameter extraction strategy for the BSIM-CMG model at room temperature. It can be divided into the following steps:

1. The first step in the parameter extraction process is to fix the values of the parameters which are directly measured or specified by the user. These parameters are the device geometry, the model selector switches, and the parameters which can be taken directly from the technology information. A list of such parameters in the BSIM-CMG model [9] is given in Table 11.1. Here, the parameters with suffix MOD are model selector switches. These parameters are used in the model either to turn on/off a specific effect such as GIDLMOD or to invoke a different set of equations to model the same effect like RDSMOD. It is good practice to turn off a specific effect via these model selector parameters if the effect is not present in the data or is not important for the final application in which the model will be used. This avoids unnecessary model equation evaluations and improves model speed.

2. After the above-mentioned parameters have been set to appropriate values, the global parameters should be initialized. This step is indicated as the parameter initialization step in Figure 11.1. It consist of three substeps, each setting specific parameters. The following trends in the measured data should be observed for this:

Table 11.1 Parameters that Should Be Set Before Starting the Extraction Flow

Parameter Name	Description
EOT	Gate oxide thickness
HFIN	Fin height
TFIN	Fin thickness
L	Fin length drawn
NFIN	Number of fins
NF	Number of fingers in parallel
NBODY	Channel doping concentration
BULKMOD	0: SOI; 1: bulk
GIDLMOD	0: off; I: on
GEOMOD	0: double gate; 1: triple gate; 2: quadruple gate
RDSMOD	0: internal; 1: external
DEVTYPE	0: PMOS; I: NMOS
NGATE	0: metal gate; >0: poly-gate doping

2.1 Plot the resistance $R_d = V_d(\sim 0.05\ \text{V})/I_d(V_g)$ versus L. Make a linear fit to this curve. Extrapolate each straight line and find the intersection point. The lines intersect at $(\Delta L, R_{series})$. The values of parameters LINT $= \Delta L/2$ and RDSW $= R_{series}$ are thus obtained. An example of such a curve if given in Figure 11.2. This step is becoming increasingly important in scaled technology owing to the high R_{series} as compared with the intrinsic channel resistance. It is possible and in fact quite common for ΔL to be negative in

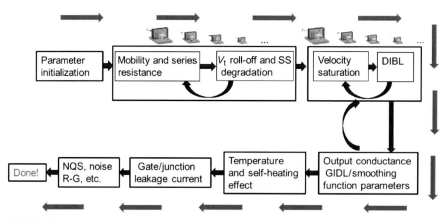

FIGURE 11.1

Parameter extraction flowchart for the BSIM-CMG model.

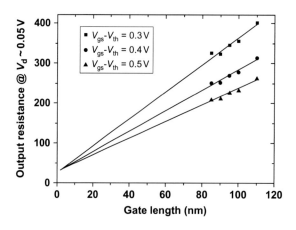

FIGURE 11.2

Plot showing $V_{ds}(\sim 0.05\ \text{V})/I_{ds}$ versus L for different values of V_g. The lines intersect at $(\Delta L, R_{series})$.

aggressive production technologies. While not intuitive, this is true because of the light doping in the source and drain just outside the gate edges and the small gate oxide thickness. The gate voltage has a stronger effect on the conductance of this region than the impurity doping. According to MOSFET theory, the channel conductance is controlled by the gate voltage, while the source drain conductance is constant. Therefore, this region ought to belong to the channel more than the source drain. A compact model such as BSIM-CMG not only captures this effect but also provides additional parameters to describe the small V_g dependence of the resistance of the source drain (region outside $L + \Delta L$) as shown in Section 4.12 of Chapter 4 and 7.2 of Chapter 7.

2.2 Plot the threshold voltage (V_{th}) difference $\Delta V_{th} = V_{th(short)} - V_{th(long)}$ versus L for $V_d \sim 0.05$ V and $V_d = V_{dd}$. V_{th} can be extracted from either a constant-current or maximum-slope extrapolation algorithm. From this plot, parameters associated with short-channel effects (SCE) and reverse SCE (RSCE) can be extracted. These parameters are DVT0, DVT1, ETA0, DSUB, K1RSCE, and LPE0. An example plot is shown in Figure 11.3. SCE are observed as the decrease of V_{th} for short-channel-length devices, while the opposite is true for RSCE. Usually an increase in V_{th} with

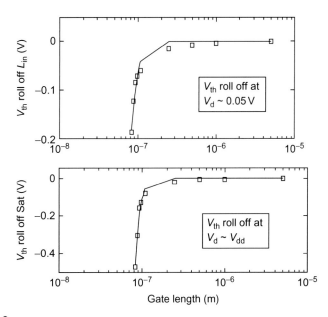

FIGURE 11.3

V_{th} roll-off curves for linear and saturation drain-voltage conditions are used to extract SCE and RSCE model parameters.

decreasing channel length (RSCE) is seen only for linear V_{th} ($V_d \sim 0.05$ V), while the saturation condition V_{th} ($V_d = V_{dd}$) decreases with channel-length reduction. This different behavior in linear and saturation condition V_{th} can be modeled in BSIM-CMG by appropriately tuning the RSCE and drain-induced barrier lowering (DIBL) parameters.

2.3 Plot the subthreshold slope SS = $(\text{d} \log_{10} I_d / \text{d} V_g)^{-1}$ versus L for both the linear condition and the saturation condition. The subthreshold slope parameters CDSC, CDSCD, and DVT1SS can be extracted from this plot. An example plot is shown in Figure 11.4. Typically, SS increases as channel length is reduced, indicating a poorer gate control for short-channel devices. This behavior can be accurately modeled via the SS parameters in the BSIM-CMG model. The adjustment of SS parameters will change the V_{th} versus L fits obtained in the previous step. It is good practice to keep an eye on the V_{th} fits too while adjusting the SS parameters.

Step 2 is useful not only to initialize the values of the global parameters, but also to remove outliers from the measured data. If a device is seen to be too far from the generally observed trends when plotting V_{th}, SS, and R_{ds} versus L, it is possible that the device is an outlier. An outlier device can arise because of incorrect measurements or fabrication issues. The model parameters should not be adjusted to fit this device because it is not a representative device for this geometry.

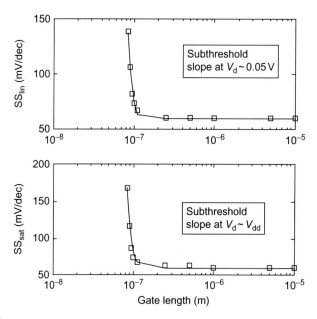

FIGURE 11.4

SS versus L for linear and saturation conditions are used to extract CDSC, CDSCD, and DVT1SS.

3. Once the scaling parameters have been set to reasonable values, the next step focuses on the fitting of linear $I_d - V_g$ curves for long- and short-channel devices. It is divided into the following substeps:

   3.1 First the work function, interface charge, and mobility parameters are extracted for a long-channel device. The work-function parameter PHIG and the interface charge parameter CIT are extracted by fitting the threshold voltage and the subthreshold slope of the $I_d - V_g$ curve, while the mobility parameters are extracted from the $I_d - V_g$ curve at V_g values above V_{th}. The $G_m - V_g$ curve is also very useful in extracting these parameters, especially the mobility degradation parameters such as UA, UD, EU, and ETAMOB. Table 11.2 shows the model parameters along with the experimental data from which they should be extracted in this step. Note that $U0_0$, UA_0, and UD_0 are the global parameters present in the length-scaling equations for local parameters U0, UA, and UD, respectively.

   3.2 Extraction of the above-mentioned parameters can slightly change the V_{th} versus L and SS versus L behavior of the model in the linear condition. Therefore, V_{th} roll-off and the SS scaling parameters set in step 2 should be fine-tuned. Table 11.3 shows the parameters which should be fine-tuned at this stage.

   3.3 Next, the low-field-mobility scaling parameters can be extracted from the $I_d - V_g$ characteristics of the long-channel and the medium-channel-length devices. The parameter names and extraction method are indicated in Table 11.4.

Table 11.2 Long-Channel Work Function, Interface Charge, and Mobility Parameter Names and the Experimental Data from Which They Should Be Extracted

Extracted Parameters	Device and Experimental Data	Extraction Methodology
PHIG, CIT	A long device I_d versus V_g @ $V_d \sim 0.05$ V	Observe subthreshold region offset and slope
$U0_0$, UA_0, UD_0, EU, ETAMOB	A long device I_d versus V_g @ $V_d \sim 0.05$ V	Observe strong inversion region Idlin and $G_m$lin

Table 11.3 Model Parameters Which Should Be Fine-Tuned in Step 3.2

Extracted Parameters	Device and Experimental Data	Extraction Methodology
DVT0, DVT1, CDSC, DVT2	Both short and medium devices I_d versus V_g @ $V_d \sim 0.05$ V	Observe subthreshold region of all devices in the same plot. Optimize DVT0, DVT1, CDSC, DVT2

Table 11.4 Extraction of the Low-Field-Mobility Scaling Parameters for Long-Channel and Medium-Channel-Length Devices

Extracted Parameters	Device and Experimental Data	Extraction Methodology
UP, LPA	Long and medium devices I_d versus V_g @ $V_d \sim 0.05$ V $U_0[L] = U0_0 \times (1 - UP \times L_{eff}^{-LPA})$	Observe strong inversion region Idlin and $G_m$lin, extract U0[L] to get UP, LP, i.e., for each L_i

Up to this stage, the linear-condition subthreshold and strong-inversion current for long-channel and medium-channel-length devices is fitted to the data. Low-field mobility is extracted from a long-channel device since short-channel devices suffer from series resistance and enhanced mobility degradation effects.

3.4 After step 3.3, short-channel device parameters for the linear condition can be extracted. For this, linear $I_d - V_g$ and $G_m - V_g$ for short-channel and medium-channel-length devices should be plotted. Observe the values of UA(L), UD(L), RDSW(L), and $\Delta L(L)$ required to fit the various channel lengths. The L dependence seen for UA, UD, RDSW, and ΔL can then be fitted using the global parameters AUA, BUA, AUB, BUB, ARDSW, BRDSW, LL, and LLN simultaneously. Typically, L dependence of ΔL is not required, and it is included in the model for additional flexibility. This step may also require fine-tuning of the low-field-mobility scaling parameters UA and LPA. An example plot of the channel-length dependence of low-field mobility is shown in Figure 11.5.

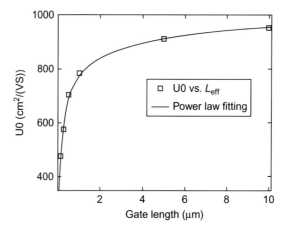

FIGURE 11.5

Low-field mobility versus channel length.

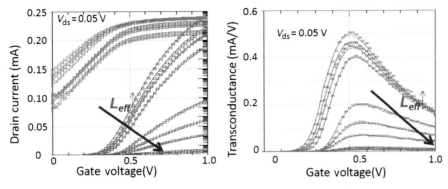

FIGURE 11.6

Sample fitting results obtained with the BSIM-CMG model after extraction of linear $I_d - V_g$ parameters for devices with channel lengths varying from 10 μm to 30 nm.

This completes step 3, which consisted of fitting linear $I_d - V_g$ from long-channel to short-channel devices. A sample fitting result obtained after this step is shown in Figure 11.6.

4. After the linear $I_d - V_g$ characteristics have been fitted, the saturation $I_d - V_g$ can be fitted. Similarly to step 3, this step is also be divided into various substeps:

   4.1 First, the DIBL parameters for long-channel and medium-channel-length devices should be optimized. Table 11.5 shows the parameters and indicates the data which should be used for their extraction. CDSCD can be tuned to fit SS in the saturation condition without a significant impact on the linear SS. ETA0 can be extracted from fitting the saturation V_{th}.

   4.2 The velocity saturation parameters for long-channel and medium-channel-length devices should be extracted after optimizing the DIBL parameters. These parameters are shown in Table 11.6. Here, PARAM0 (e.g., VSAT0 in Table 11.6) is the global parameter modeling the geometry dependence of the local parameter PARAM (e.g., VSAT in Table 11.6). VSAT1 is used in the output conductance model, while VSAT is used in the calculation of saturation voltage V_{dsat} (see Sections 4.6 and 4.7 of Chapter 4). Typically, VSAT should be equal to VSAT1, and they are

Table 11.5 DIBL Effect Parameters and the Experimental Data from Which They Should Be Extracted

Extracted Parameters	Device and Experimental Data	Extraction Methodology
ETA0, DSUB, CDSCD	Short and long devices I_d versus V_g @ $V_d \sim V_{dd}$	Observe subthreshold region of all devices in the same plot. Optimize ETA0, DSUB, CDSCD

Table 11.6 Velocity Saturation Parameters for Long-Channel and Medium-Channel-Length Devices

Extracted Parameters	Device and Experimental Data	Extraction Methodology
$VSAT_0$, $VSAT1_0$, $PTWG_0$, $KSATIV_0$, $MEXP_0$	Long device and medium devices I_d versus V_g @ $V_d \sim V_{dd}$	Observe strong inversion region $I_d$sat, $G_m$sat, $I_d V_d$

provided separately in the BSIM-CMG model for additional flexibility in fitting. PTWG is an empirical parameter introduced in the BSIM-CMG model to improve the model fitting of saturation $G_m - V_g$ characteristics.

4.3 After the fitting for long-channel and medium-channel-length devices, the velocity saturation parameters for short-channel devices should be extracted. The length scaling of the velocity saturation parameters should be used for this purpose. The parameter names and the extraction method are given in Table 11.7.

A sample model result obtained after this step is shown in Figure 11.7 on a linear and a semilog scale.

5. Once the saturation $I_d - V_g$ characteristics have been fitted, the output characteristics should be looked at. For a long-channel device, the output characteristics should already fit well to the measured data. A fine-tuning of the long-channel output characteristics can be done using the parameter MEXP, which models the transition between linear and saturation regions. For short-channel devices, the output conductance parameters should be tuned as shown in Table 11.8. The output conductance parameters are for channel-length modulation, DIBL, and SCE. Each of these effects impacts in specific regions of the $I_d - V_d$ characteristics as discussed in Section 4.11 of Chapter 4. Accurate modeling of output conductance is facilitated by the model parameters CLM,

Table 11.7 Velocity Saturation Parameters for Short-Channel Devices

Extracted Parameters	Device & Experimental Data	Extraction Methodology
AYSAT, AVSAT1, APTWG, BVSAT, BVSAT1, BPTWG	Short and medium devices I_d versus V_g @ $V_d \sim V_{dd}$	a. Observe strong inversion region of $I_d$sat and $G_m$sat. Find $VSAT1[L_i] = X_i$, $VSAT[L_i] = Y_i$, $PTWG[L_i] = Z_i$ to fit data b. Extract AVSATI, BVSATI from (L_i, X_i); AVSAT.BVSAT from (L_i, Y_i); APTWG, BPTWG from (L_i, Z_i)

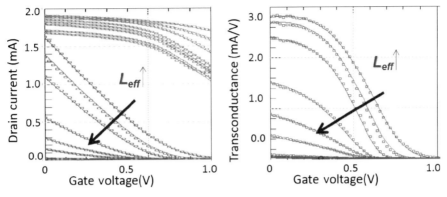

FIGURE 11.7

Fitting results obtained for saturation $I_d - V_g$ characteristics with the BSIM-CMG model for PMOS devices with channel lengths varying from 10 μm to 30 nm.

Table 11.8 BSIM-CMG Parameters for Fitting the Output Characteristics

Extracted Parameters	Device and Experimental Data	Extraction Methodology
MEXP[L], PCLM, PDIBL1, PDIBL2, DROUT, PVAG	Long and short devices I_d versus V_d @ differently V_g	Observe strong inversion region I_d versus V_d and G_d versus V_d @ different V_g

DIBL, and SCE. For further improvement in the fitting, the L dependence of MEXP can be used. A sample fitting for $I_d - V_d$ and $G_d - V_d$ after this step is shown in Figure 11.8 for an $L = 90$ nm n-type MOS device.

6. Once the drain-current model parameters for all the device have been extracted, the parameters for additional effects such as gate-induced drain leakage (GIDL), gate leakage, and temperature dependence can be extracted from the region of the device characteristic they dominate. Gate-leakage current parameter extraction may change the fitting of large-area (wide long-channel) devices slightly, while the GIDL model parameters may slightly affect the off-state current predicted by the model. The temperature dependence of device characteristics is modeled via the temperature dependence of key model parameters such mobility and threshold voltage (see Chapter 12). The temperature parameters can be easily extracted from device characteristics measured at multiple ambient temperatures.

7. This completes the guideline steps for extraction for BSIM-CMG drain-current model parameters. It was shown that the BSIM-CMG model parameters can be extracted systematically to fit devices with channel lengths ranging from long to very short. More advanced model extraction such as extraction of RF model parameters can follow the steps indicated in [9] for the BSIM6 model.

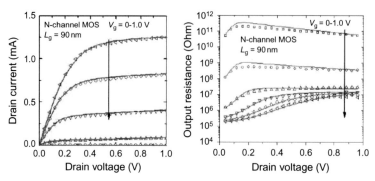

FIGURE 11.8

Fitting results obtained for $I_d - V_d$ and $G_d - V_d$ using the BSIM-CMG model.

11.3 CONCLUSION

In conclusion, parameter extraction steps for the BSIM-CMG model were discussed in this chapter. These steps are guideline steps, and there can be variations on these depending on the availability or unavailability of certain measurement data. The significance of modeling a specific region of the device characteristics also depends on the final application of the model. This may also alter the parameter extraction flow. In any case, after parameter extraction, it is important to do a sanity check on the extracted values. The parameter values should lie in a sensible range and scale with the device geometry and ambient temperature in a reasonable way. It is good practice to run the benchmark tests discussed in Chapter 10 on the model card obtained after the parameter extraction.

REFERENCES

[1] K. Levenberg, A method for the solution of certain non-linear problems in least squares, Q. J. Appl. Math. II (1944) 164–168.

[2] D.E. Goldberg, Genetic Algorithm in Search, Optimization and Machine Learning, Addison-Wesley Professional, Boston, MA, 1989.

[3] A.M. Chopde, S. Khandelwal, R.A. Thakker, M.B. Patil, K.G. Anil, Parameter extraction for MOS Model 11 using particle swarm optimization, Proceedings of the International Workshop on Physics of Semiconductor Devices, 2007, pp. 253–256.

[4] J. Watts, C. Bittner, D. Heaberlin, J. Hoffman, Extraction of compact model parameters for ULSI MOSFETs using a genetic algorithm, Proceedings of the International Conference on Modeling and Simulation of Microsystems, November 1999, pp. 176–179.

[5] BSIMProPlus, Proplus, http://www.proplussolutions.com/en/pro1/Advanced-SPICE-Modeling-Platform---BSIMProPlus.html.

[6] ICCAP Users' Manual, Agilent Technologies, Available: http://cp.literature.agilent.com/litweb/pdf/iccap2008/iccap2008.html.

[7] Model Parameter Builder Users' Manual, Accelicon (now Agilent), Available: http://www.home.agilent.com/en/pc-2112961/model-builder-program-silicon-focused-device-modeling-software?nid=33185.0.00&cc=US&lc=eng.

[8] UTMOST Users' Manual, Silvaco, Available: http://www.silvaco.com/products/analog_mixed_signal/device_characterization_modeling/utmost_III.html.

[9] BSIM-CMG Users' Manual, BSIM Group. Available: http://www-device.eecs.berkeley.edu/bsim/.

[10] S. Yao, T. Morshed, D.D. Lu, S. Venugopalan, W. Xiong, C.R. Cleaelin, A. M. Niknejad, C. Hu, Global parameter extraction for a multi-gate MOSFET compact model, Proc. ICMTS March 2010, pp. 9.1–9.4.

[11] S. Venugopalan, K. Dandu, S. Martin, R. Taylor, C. Cirba, X. Zhang, A.M. Niknejad, C. Hu, A non-iterative physical procedure for RF CMOS compact model extraction using BSIM6, Proc. CICC 2012, pp. 1–4.

Temperature dependence

12

CHAPTER OUTLINE

12.1 Semiconductor properties... 246
    12.1.1 Band gap temperature dependence 246
    12.1.2 Temperature dependence of N_c, V_{bi}, and Φ_B...................... 246
    12.1.3 Temperature dependence of the intrinsic carrier concentration 247
12.2 Temperature dependence of the threshold voltage.............................. 247
    12.2.1 Temperature dependence of drain-induced barrier lowering........ 248
    12.2.2 Temperature dependence of the body effect 248
    12.2.3 Subthreshold swing 248
12.3 Temperature dependence of mobility 249
12.4 Temperature dependence of velocity saturation 249
    12.4.1 Temperature dependence of the nonsaturation effect 250
12.5 Temperature dependence of leakage currents 250
    12.5.1 Gate current 250
    12.5.2 Gate-induced drain/source leakage 250
    12.5.3 Impact ionization 251
12.6 Temperature dependence of parasitic source/drain resistances................. 251
12.7 Temperature dependence of source/drain diode characteristics 252
    12.7.1 Direct current model 252
    12.7.2 Capacitance 254
    12.7.3 Trap-assisted tunneling current 254
12.8 Self-heating effect 256
12.9 Validation range... 257
12.10 Model validation on measured data... 257
References... 260

It is well known that any change in operating temperature will affect the device characteristics and hence change circuit performance. In fact, the major reason for not increasing the operating frequency of microprocessors is the power density, which leads to a huge increase in temperature, thereby affecting performance and reliability. For any model, it is essential to provide accurate device characteristics for the entire temperature range. Ideally, the model should be able to predict this behavior for all temperatures, but this is not possible practically. Normally most compact MOSFET

FinFET Modeling for IC Simulation and Design. http://dx.doi.org/10.1016/B978-0-12-420031-9.00012-9

models are valid and tested for -50 to $150\,°C$, which is also the normal operation range of most of the circuits. An accurate temperature model requires accurate modeling of the temperature dependence of the different device parameters, such as threshold voltage, mobility, parasitic source/drain resistances, and source-drain junction parameters. Here, we briefly describe the temperature dependence models of BSIM-CMG.

The nominal parameter extraction is assumed to have been performed at TNOM, which is a model parameter. Generally, TNOM is taken as $300.15\,K$ as nominal extraction is done at room temperature.

12.1 SEMICONDUCTOR PROPERTIES

The basic properties of intrinsic and extrinsic semiconductors change with temperature and are discussed in detail in device physics books [1]. Here, we will discuss a few of these which involve model parameters as they may need to be modified during parameter extraction.

12.1.1 BAND GAP TEMPERATURE DEPENDENCE

The band gap of a semiconductor decreases with an increase in temperature [2]. The spacing between atoms increases when the amplitude of the atomic vibrations increases owing to the increased thermal energy. The increased interatomic spacing decreases the potential seen by the electrons in the semiconductor, which in turn reduces the size of the energy band gap [3]. The temperature dependence of the band gap is modeled empirically in the literature using Varshni's expression [4]:

$$E_g = \text{BG0SUB} - \frac{\text{TBGASUB} \cdot T^2}{T + \text{TBGBSUB}}, \tag{12.1}$$

where T is the temperature in kelvins, and BG0SUB, TBGASUB, and TBGBSUB are the model parameters.

The direct modulation of the interatomic distance, such as by applying high compressive (tensile) stress, also causes an increase (decrease) of the band gap. This principle is used in all modern integrated circuits—strained silicon transistor.

12.1.2 TEMPERATURE DEPENDENCE OF N_C, V_{bi}, AND Φ_B

The temperature dependence of the effective density of states, source-drain built-in potential, and Φ_B are given by

$$N_C = \text{NC0SUB} \cdot \left(\frac{T}{300.15}\right)^{3/2}, \tag{12.2}$$

$$V_{bi} = \frac{kT}{q} \cdot \ln\left(\frac{NSD \cdot NBODY}{n_i^2}\right), \tag{12.3}$$

$$\Phi_B = \frac{kT}{q} \cdot \ln\left(\frac{NBODY}{n_i}\right). \tag{12.4}$$

12.1.3 TEMPERATURE DEPENDENCE OF THE INTRINSIC CARRIER CONCENTRATION

The intrinsic carrier concentration is expressed as

$$n_i = \sqrt{N_C N_V} \cdot \exp\left(-\frac{E_g}{2kT}\right), \tag{12.5}$$

where N_C and N_V are the effective density of states in the conduction and valence bands, respectively, and are a function of temperature (proportional to $T^{3/2}$). The main temperature dependence of n_i comes from the exponential part, which causes n_i to increase with an increase in temperature. This is why a semiconductor becomes "intrinsic" at very high temperatures as n_i can become larger than the dopant concentration. At the other extreme (at very low temperatures), there may be a freeze-out effect, which causes dopant nonionization. This is why undoped group III-V semiconductor materials are preferred over silicon at cryogenic temperatures to avoid freeze-out.

Taking out the temperature dependence from Equation (12.5) and denoting n_i as parameter NI0SUB at $T = $ TNOM, we can write

$$n_i = NI0SUB \cdot \left(\frac{T}{TNOM}\right)^{3/2} \cdot \exp\left(\frac{BG0SUB}{2k \cdot TNOM} - \frac{E_g}{2k \cdot T}\right), \tag{12.6}$$

where BG0SUB is a parameter which refers to the band gap at $T = $ TNOM.

12.2 TEMPERATURE DEPENDENCE OF THE THRESHOLD VOLTAGE

The threshold voltage (V_{th}) decreases with an increase in temperature owing to a shift in the Fermi energy level and a decrease in the band gap. Although the temperature dependence of semiconductor properties is captured through their parameters as described earlier, we do not use it for the threshold voltage. The main reason for this approach is to decouple different effects and thereby parameters to provide flexibility in model fitting.

This dependence of the threshold voltage is almost linear for long-channel devices [5] for a wide range of temperature and is modeled as [6–8]

$$V_{th}(T) = V_{th}(TNOM, L, V_{ds}) + \left(KT1 + \frac{KT1L1}{L}\right)\left(\frac{T}{TNOM} - 1\right). \tag{12.7}$$

The term KT1L$/L$ is introduced to aid in better fitting. The temperature dependence of the band gap E_g, the surface potential ϕ_s at threshold, and the intrinsic carrier concentration n_i are all accounted for by KT1, which is represented by a simple linear equation. This is done to simplify parameter extraction and make the model easier to fit. The temperature dependence of E_g and n_i is maintained while evaluating the saturation current of the source-drain junction.

12.2.1 TEMPERATURE DEPENDENCE OF DRAIN-INDUCED BARRIER LOWERING

Normally, the drain voltage dependence of the threshold voltage—that is, drain-induced barrier lowering in a MOSFET, is independent of temperature [6]. However, to provide flexibility for better fitting across drain bias especially for short-channel devices, the parameter TETA0 is provided:

$$\text{ETA0}(T) = \text{ETA0} \cdot [1 - \text{TETA0} \cdot (T - \text{TNOM})], \tag{12.8}$$

$$\text{ETA0R}(T) = \text{ETA0R} \cdot [1 - \text{TETA0R} \cdot (T - \text{TNOM})]. \tag{12.9}$$

The parameter TETA0R can be used when $\text{ASYMMOD} = 1$.

12.2.2 TEMPERATURE DEPENDENCE OF THE BODY EFFECT

The substrate bias effect may also have a temperature dependence, which is captured by

$$\text{K0}(T) = \text{K0} + \text{K01} \cdot (T - \text{TNOM}), \tag{12.10}$$

$$\text{K1}(T) = \text{K1} + \text{K11} \cdot (T - \text{TNOM}), \tag{12.11}$$

$$\text{K0SI}(T) = \text{K0SI} + \text{K0SI1} \cdot (T - \text{TNOM}), \tag{12.12}$$

$$\text{K1SI}(T) = \text{K1SI} + \text{KISI1} \cdot (T - \text{TNOM}), \tag{12.13}$$

$$\text{KISAT}(T) = \text{KISAT} + \text{KISAT1} \cdot (T - \text{TNOM}). \tag{12.14}$$

12.2.3 SUBTHRESHOLD SWING

The subthreshold swing is very sensitive to temperature as diffusion current dominates in this region of operation. The off current increases exponentially with a decrease in temperature. The subthreshold slope varies almost linearly with temperature and is modeled as

$$\Theta_{SS} = 1 + \text{TSS} \cdot (T - \text{TNOM}). \tag{12.15}$$

The temperature-dependent subthreshold factor Θ_{SS} is multiplied to n in (4.31), which is then used in the surface potential calculation.

12.3 **TEMPERATURE DEPENDENCE OF MOBILITY**

There has been lot of research on accurate modeling of the temperature dependence of mobility [9, 10]. The characteristics of carrier mobility in the inversion layer are governed by three major scattering mechanisms: phonon scattering, surface scattering, and coulomb scattering (including ionized impurity scattering and interface charge scattering). Each of these scattering mechanisms dominates over the others in different temperature ranges [11]; for example, phonon scattering becomes the dominant mechanism at temperatures above 250 K, while coulomb scattering becomes important for very low temperatures. The temperature dependence of low-field mobility is modeled as

$$\mu_0(T) = U0 \cdot \left(\frac{T}{\text{TNOM}} \right)^{\text{UTE}} + \text{UTL} \cdot (T - \text{TNOM}),$$ (12.16)

where T is the temperature in kelvins and TNOM is the temperature at which the nominal model parameters are extracted. In the mobility model, ETAMOB, UA, UC, UD, and UCS also have temperature dependence and are modeled by the following equations:

$$\text{ETAMOB}(T) = \text{ETAMOB} \cdot [1 + \text{EMOBT} \cdot (T - \text{TNOM})],$$ (12.17)

$$\text{UA}(T) = \text{UA} + \text{UA1} \cdot (T - \text{TNOM}),$$ (12.18)

$$\text{UC}(T) = \text{UC} \cdot [1 + \text{UC1} \cdot (T - \text{TNOM})],$$ (12.19)

$$\text{UD}(T) = \text{UD} \cdot \left(\frac{T}{\text{TNOM}} \right)^{\text{UD1}},$$ (12.20)

$$\text{UCS}(T) = \text{UCS} \cdot \left(\frac{T}{\text{TNOM}} \right)^{\text{UCSTE}},$$ (12.21)

The parameters U0, ETAMOB, UA, UC, UD, and UCS are extracted at nominal temperature.

12.4 **TEMPERATURE DEPENDENCE OF VELOCITY SATURATION**

The carrier velocity saturates at high field as discussed and modeled in Chapter 6. The electron velocity in the inversion layer decreases with an increase in temperature owing to increased scattering caused by lattice vibrations. The temperature dependence of the saturation velocity (v_{SAT}) in the inversion layer is linear as demonstrated experimentally in the literature [6–8], and is modeled as follows in BSIM-CMG:

$$\text{VSAT}(T) = \text{VSAT} \cdot [1 - \text{AT} \cdot (T - \text{TNOM})],$$ (12.22)

$$\text{VSAT1}(T) = \text{VSAT1} \cdot [1 - \text{AT} \cdot (T - \text{TNOM})],$$ (12.23)

$$\text{VSAT1R}(T) = \text{VSAT1R} \cdot [1 - \text{AT} \cdot (T - \text{TNOM})],$$ (12.24)

$$\text{VSATCV}(T) = \text{VSATCV} \cdot [1 - \text{AT} \cdot (T - \text{TNOM})], \qquad (12.25)$$

$$\text{PTWG}(T) = \text{PTWG} \cdot [1 - \text{PTWGT} \cdot (T - \text{TNOM})], \qquad (12.26)$$

$$\text{PTWGR}(T) = \text{PTWGR} \cdot [1 - \text{PTWGT} \cdot (T - \text{TNOM})], \qquad (12.27)$$

$$\text{MEXP}(T) = \text{MEXP} \cdot [1 + \text{TMEXP} \cdot (T - \text{TNOM})], \qquad (12.28)$$

The saturation velocity is a weak function of temperature in contrast to the other quantities in a MOSFET. This is why, a MOSFET will not provide significant gain when used at low temperatures as the saturation current does not increase significantly.

12.4.1 TEMPERATURE DEPENDENCE OF THE NONSATURATION EFFECT

The temperature dependence of the nonsaturation effect (Equation 4.71) is also modeled linearly as follows:

$$\text{A1}(T) = \text{A1} + \text{A11} \cdot (T - \text{TNOM}), \qquad (12.29)$$

$$\text{A2}(T) = \text{A2} + \text{A21} \cdot (T - \text{TNOM}). \qquad (12.30)$$

12.5 TEMPERATURE DEPENDENCE OF LEAKAGE CURRENTS

The temperature dependence of leakage currents is modeled empirically as follows:

12.5.1 GATE CURRENT

$$\text{AIGBINV}(T) = \text{AIGBINV} + \text{AIGBINV1} \cdot (T - \text{TNOM}), \qquad (12.31)$$

$$\text{AIGBACC}(T) = \text{AIGBACC} + \text{AIGBACC1} \cdot (T - \text{TNOM}), \qquad (12.32)$$

$$\text{AIGC}(T) = \text{AIGC} + \text{AIGC1} \cdot (T - \text{TNOM}), \qquad (12.33)$$

$$\text{AIGS}(T) = \text{AIGS} + \text{AIGS1} \cdot (T - \text{TNOM}), \qquad (12.34)$$

$$\text{AIGD}(T) = \text{AIGD} + \text{AIGD1} \cdot (T - \text{TNOM}), \qquad (12.35)$$

$$\text{igtemp} = \left(\frac{T}{\text{TNOM}} \right)^{\text{IGT}}. \qquad (12.36)$$

This igtemp is multiplied by the gate current equation.

12.5.2 GATE-INDUCED DRAIN/SOURCE LEAKAGE

$$\text{BGIDL}(T) = \text{BGIDL} \cdot [1 + \text{TGIDL} \cdot (T - \text{TNOM})], \qquad (12.37)$$

$$\text{BGISL}(T) = \text{BGISL} \cdot [1 + \text{TGIDL} \cdot (T - \text{TNOM})]. \qquad (12.38)$$

12.5.3 IMPACT IONIZATION

$$\text{ALPHA0}(T) = \text{ALPHA0} + \text{ALPHA01} \cdot (T - \text{TNOM}), \tag{12.39}$$

$$\text{ALPHA1}(T) = \text{ALPHA1} + \text{ALPHA11} \cdot (T - \text{TNOM}), \tag{12.40}$$

$$\text{ALPHAII0}(T) = \text{ALPHAII0} + \text{ALPHAII01} \cdot (T - \text{TNOM}), \tag{12.41}$$

$$\text{ALPHAII1}(T) = \text{ALPHAII1} + \text{ALPHAII11} \cdot (T - \text{TNOM}), \tag{12.42}$$

$$\text{BETA0}(T) = \text{BETA0} \cdot \left(\frac{T}{\text{TNOM}} \right)^{\text{IIT}}, \tag{12.43}$$

$$\text{SII0}(T) = \text{SII0} \left[1 + \text{TII} \left(\frac{T}{\text{TNOM}} - 1 \right) \right]. \tag{12.44}$$

12.6 TEMPERATURE DEPENDENCE OF PARASITIC SOURCE/DRAIN RESISTANCES

For FinFETs or in general any short-channel device, parasitic source/drain resistances significantly decrease the drive current. The accurate modeling of these parasitic resistances has already been discussed in detail in Chapter 7. The contact resistance, diffusion resistance, and spreading resistance at the edge of the inversion layer may have different temperature coefficients because of different materials and/or doping combinations. All of these resistances are taken into account in the model for R_{ds}, which increases linearly with increasing temperature [6, 8] and is modeled as follows,

For RDSMOD $= 1$ and ASYMOD $= 0$,

$$\text{RSWMIN}(T) = \text{RSWMIN} \cdot [1 + \text{PRT} \cdot (T - \text{TNOM})], \tag{12.45}$$

$$\text{RDWMIN}(T) = \text{RDWMIN} \cdot [1 + \text{PRT} \cdot (T - \text{TNOM})], \tag{12.46}$$

$$\text{RSW}(T) = \text{RSW} \cdot [1 + \text{PRT} \cdot (T - \text{TNOM})], \tag{12.47}$$

$$\text{RDW}(T) = \text{RDW} \cdot [1 + \text{PRT} \cdot (T - \text{TNOM})]. \tag{12.48}$$

For RDSMOD $= 0$ and 2,

$$\text{RDSWMIN}(T) = \text{RDSWMIN} \cdot [1 + \text{PRT} \cdot (T - \text{TNOM})], \tag{12.49}$$

$$\text{RDSW}(T) = \text{RDSW} \cdot [1 + \text{PRT} \cdot (T - \text{TNOM})], \tag{12.50}$$

$$\text{RSWMIN}(T) = \text{RSWMIN} \cdot [1 + \text{PRT} \cdot (T - \text{TNOM})], \tag{12.51}$$

$$\text{RDWMIN}(T) = \text{RDWMIN} \cdot [1 + \text{PRT} \cdot (T - \text{TNOM})]. \tag{12.52}$$

For RDSMOD $= 2$,

$$R_{\text{s,geo}}(T) = R_{\text{s,geo}} \cdot [1 + \text{PRT} \cdot (T - \text{TNOM})], \tag{12.53}$$

$$R_{\text{d,geo}}(T) = R_{\text{d,geo}} \cdot [1 + \text{PRT} \cdot (T - \text{TNOM})]. \tag{12.54}$$

For RDSMOD $= 1$ and ASYMOD $= 1$, the following parameters may be used to bring asymmetry to the temperature dependence of the drain-side and source-side resistances;

$$\text{RSDR}(T) = \text{RSDR} \cdot [1 + \text{TRSDR} \cdot (T - \text{TNOM})], \quad (12.55)$$

$$\text{RSDRR}(T) = \text{RSDRR} \cdot [1 + \text{TRSDR} \cdot (T - \text{TNOM})], \quad (12.56)$$

$$\text{RDDR}(T) = \text{RDDR} \cdot [1 + \text{TRDDR} \cdot (T - \text{TNOM})], \quad (12.57)$$

$$\text{RDDRR}(T) = \text{RDDRR} \cdot [1 + \text{TRDDR} \cdot (T - \text{TNOM})]. \quad (12.58)$$

12.7 TEMPERATURE DEPENDENCE OF SOURCE/DRAIN DIODE CHARACTERISTICS

The p-n junction diode has temperature dependence in both of its components—that is, current and capacitance. Both of these need to be modeled at the zero-bias condition. The temperature dependence of direct current comes from the saturation current, which in turn gets it from the temperature dependence of the intrinsic carrier density (n_i) and the energy band gap (E_g) [5, 12, 13]. The temperature dependence of the junction capacitance depends on the temperature dependence of the dielectric constant of silicon and the junction built-in potential [5, 13].

12.7.1 DIRECT CURRENT MODEL

The saturation current comprises the generation current caused by thermal generation of electron-hole pairs inside the depletion region and the diffusion current due to minority carriers diffusing across the depletion region [1]. Both of these currents are a strong function of temperature as follows:

$$I_{\text{generation}} \propto n_i \propto T^{3/2} \exp\left(-\frac{E_g}{2kT}\right), \quad (12.59)$$

$$I_{\text{diffussion}} \propto n_i{}^2 \propto T^3 \exp\left(-\frac{E_g}{kT}\right). \quad (12.60)$$

As explained in Chapter 9, both of these are combined in a single current equation as shown below (parameter names used from source-side junctions):

$$I_{\text{saturation}} \propto T^{\text{XTIS}} \exp\left(-\frac{E_g}{\text{NJS} \cdot kT}\right), \quad (12.61)$$

or

$$I_{\text{saturation}} \propto \exp\left(\ln T^{\text{XTIS}}\right) \exp\left(-\frac{E_g}{\text{NJS} \cdot kT}\right), \quad (12.62)$$

or

$$I_{\text{saturation}} \propto \exp\left(\frac{-\frac{E_g}{kT} + \text{XTIS} \cdot \ln T}{\text{NJS}}\right). \tag{12.63}$$

The parameter XTIS is a junction current temperature exponent and NJS is a junctions emission coefficient.

The source-side and drain-side junction saturation currents are modeled as follows.

The source-side junction saturation current is given by

$$I_{\text{sbs}} = J_{\text{ss}}(T) \cdot \text{ASEJ} + J_{\text{sws}}(T) \cdot \text{PSEJ} + J_{\text{swgs}}(T) \cdot W_{\text{eff0}} \cdot \text{NFIN}_{\text{total}}, \tag{12.64}$$

where ASEJ is the area of the bottom of the source and PSEJ is the perimeter, J_{ss} is the saturation current density (per unit area) at the bottom of the junction, J_{sws} is the saturation sidewall current density (per unit length), and J_{swgs} is the gate sidewall saturation current density (per unit length). Using Equation (12.63), we can write the temperature dependence of the source-side current densities as follows:

$$J_{\text{ss}}(T) = \text{JSS} \cdot \exp\left(\frac{\frac{E_{g,\text{TNOM}}}{k\text{TNOM}} - \frac{E_g}{kT} + \text{XTIS} \cdot \ln\left(\frac{T}{\text{TNOM}}\right)}{\text{NJS}}\right), \tag{12.65}$$

$$J_{\text{sws}}(T) = \text{JSWS} \cdot \exp\left(\frac{\frac{E_{g,\text{TNOM}}}{k\text{TNOM}} - \frac{E_g}{kT} + \text{XTIS} \cdot \ln\left(\frac{T}{\text{TNOM}}\right)}{\text{NJS}}\right), \tag{12.66}$$

$$J_{\text{swgs}}(T) = \text{JSWGS} \cdot \exp\left(\frac{\frac{E_{g,\text{TNOM}}}{k\text{TNOM}} - \frac{E_g}{kT} + \text{XTIS} \cdot \ln\left(\frac{T}{\text{TNOM}}\right)}{\text{NJS}}\right). \tag{12.67}$$

Similarly, for the drain side, we have

$$I_{\text{sbd}} = J_{\text{sd}}(T) \cdot \text{ADEJ} + J_{\text{swd}}(T) \cdot \text{PDEJ} + J_{\text{swgd}}(T) \cdot W_{\text{eff0}} \cdot \text{NFIN}_{\text{total}}, \tag{12.68}$$

where ADEJ is the area of the bottom of the drain and PDEJ is the perimeter, J_{sd} is the saturation current density (per unit area) at the bottom of the junction, J_{swd} is the saturation sidewall current density (per unit length), and J_{swgd} is the gate sidewall saturation current density (per unit length). Using Equation (12.63), we can write the temperature dependence of the drain-side current densities as follows:

$$J_{\text{sd}}(T) = \text{JSD} \cdot \exp\left(\frac{\frac{E_{g,\text{TNOM}}}{k\text{TNOM}} - \frac{E_g}{kT} + \text{XTID} \cdot \ln\left(\frac{T}{\text{TNOM}}\right)}{\text{NJD}}\right), \tag{12.69}$$

$$J_{\text{swd}}(T) = \text{JSWD} \cdot \exp\left(\frac{\frac{E_{g,\text{TNOM}}}{k\text{TNOM}} - \frac{E_g}{kT} + \text{XTID} \cdot \ln\left(\frac{T}{\text{TNOM}}\right)}{\text{NJD}}\right), \tag{12.70}$$

$$J_{\text{swgd}}(T) = \text{JSWGD} \cdot \exp\left(\frac{\frac{E_{g,\text{TNOM}}}{k\text{TNOM}} - \frac{E_g}{kT} + \text{XTID} \cdot \ln\left(\frac{T}{\text{TNOM}}\right)}{\text{NJD}}\right). \tag{12.71}$$

12.7.2 CAPACITANCE

The temperature dependence of the junction capacitances is modeled through temperature-dependent zero-bias capacitances per unit area or per unit length, and temperature-dependent junction built-in potentials.

On the source side, the temperature dependence of the junction capacitances and potentials is defined as

$$\text{CJS}(T) = \text{CJS}[1 + \text{TCJ}(T - \text{TNOM})], \tag{12.72}$$

$$\text{CJSWS}(T) = \text{CJSWS}[1 + \text{TCJSW}(T - \text{TNOM})], \tag{12.73}$$

$$\text{CJSWGS}(T) = \text{CJSWGS}[1 + \text{TCJSWG}(T - \text{TNOM})], \tag{12.74}$$

$$\text{PBS}(T) = \text{PBS}(\text{TNOM}) - \text{TPB}(T - \text{TNOM}), \tag{12.75}$$

$$\text{PBSWS}(T) = \text{PBSWS}(\text{TNOM}) - \text{TPBSW}(T - \text{TNOM}), \tag{12.76}$$

$$\text{PBSWGS}(T) = \text{PBSWGS}(\text{TNOM}) - \text{TPBSWG}(T - \text{TNOM}), \tag{12.77}$$

where CJS is the junction capacitance at the bottom, CJSWS is the junction capacitance at the source sidewall, CJSWGS is the junction capacitance at the gate sidewall, PBS is the junction potential at the bottom, PBSWS is the junction capacitance at the source sidewall, and PBSWGS is the junction capacitance at the gate sidewall.

Similarly, on the drain side, we have

$$\text{CJD}(T) = \text{CJD}[1 + \text{TCJ}(T - \text{TNOM})], \tag{12.78}$$

$$\text{CJSWD}(T) = \text{CJSWD}[1 + \text{TCJSW}(T - \text{TNOM})], \tag{12.79}$$

$$\text{CJSWGD}(T) = \text{CJSWGD}[1 + \text{TCJSWG}(T - \text{TNOM})], \tag{12.80}$$

$$\text{PBD}(T) = \text{PBD}(\text{TNOM}) - \text{TPB}(T - \text{TNOM}), \tag{12.81}$$

$$\text{PBSWD}(T) = \text{PBSWD}(\text{TNOM}) - \text{TPBSW}(T - \text{TNOM}), \tag{12.82}$$

$$\text{PBSWGD}(T) = \text{PBSWGD}(\text{TNOM}) - \text{TPBSWG}(T - \text{TNOM}), \tag{12.83}$$

where CJD is the junction capacitance at the bottom, CJSWD is the junction capacitance at the source sidewall, CJSWGD is the junction capacitance at the gate sidewall, PBD is the junction potential at the bottom, PBSWD is the junction capacitance at the source sidewall, and PBSWGD is the junction capacitance at the gate sidewall.

12.7.3 TRAP-ASSISTED TUNNELING CURRENT

The temperature dependence for the trap-assisted tunneling saturation current has been modeled as follows.

On the source side, we have

$$J_{\text{tss}}(T) = \text{JTSS} \cdot \exp\left(\frac{E_{\text{g,TNOM}} \cdot \text{XTSS} \cdot \left(\frac{T}{\text{TNOM}} - 1\right)}{kT}\right), \tag{12.84}$$

$$J_{\text{tssws}}(T) = \text{JTSSWS} \cdot \exp\left(\frac{E_{\text{g,TNOM}} \cdot \text{XTSSWS} \cdot \left(\frac{T}{\text{TNOM}} - 1\right)}{kT}\right), \tag{12.85}$$

$$J_{\text{tsswgs}}(T) = \text{JTSSWGS} \times \left(\sqrt{\frac{\text{JTWEFF}}{W_{\text{eff0}}}} + 1\right) \times \exp\left(\frac{E_{\text{g,TNOM}} \cdot \text{XTSSWGS} \cdot \left(\frac{T}{\text{TNOM}} - 1\right)}{kT}\right), \tag{12.86}$$

where J_{tss} is the current density for the junction bottom, J_{tssws} is the current density for the junction sidewall, and J_{tsswgs} is the current density for the gate sidewall.

Similarly, on the drain side, we have

$$J_{\text{tsd}}(T) = \text{JTSD} \cdot \exp\left(\frac{E_{\text{g,TNOM}} \cdot \text{XTSD} \cdot \left(\frac{T}{\text{TNOM}} - 1\right)}{kT}\right), \tag{12.87}$$

$$J_{\text{tsswd}}(T) = \text{JTSSWD} \cdot \exp\left(\frac{E_{\text{g,TNOM}} \cdot \text{XTSSWD} \cdot \left(\frac{T}{\text{TNOM}} - 1\right)}{kT}\right), \tag{12.88}$$

$$J_{\text{tsswgd}}(T) = \text{JTSSWGD} \times \left(\sqrt{\frac{\text{JTWEFF}}{W_{\text{eff0}}}} + 1\right) \times \exp\left(\frac{E_{\text{g,TNOM}} \cdot \text{XTSSWGD} \cdot \left(\frac{T}{\text{TNOM}} - 1\right)}{kT}\right), \tag{12.89}$$

where J_{tsd} is current density for the junction bottom, J_{tsswd} is the current density for the junction sidewall, and J_{tsswgd} is the current density for the gate sidewall.

The body diode nonideality factors are also dependent on temperature and are modeled as follows.

For the source side, we have

$$\text{NJTS}(T) = \text{NJTS} \times \left[1 + \text{TNJTS} \cdot \left(\frac{T}{\text{TNOM}} - 1\right)\right], \tag{12.90}$$

$$\text{NJTSSW}(T) = \text{NJTSSW} \times \left[1 + \text{TNJTSSW} \cdot \left(\frac{T}{\text{TNOM}} - 1\right)\right], \tag{12.91}$$

$$\text{NJTSSWG}(T) = \text{NJTSSWG} \times \left[1 + \text{TNJTSSWG} \cdot \left(\frac{T}{\text{TNOM}} - 1\right)\right], \tag{12.92}$$

where NJTS corresponds to the junction bottom, NJTSSW corresponds to the sidewall, and NJTSSWG corresponds to the gate sidewall.

Similarly, on the drain side, we have

$$\text{NJTSD}(T) = \text{NJTSD} \times \left[1 + \text{TNJTSD} \cdot \left(\frac{T}{\text{TNOM}} - 1\right)\right], \tag{12.93}$$

$$\text{NJTSSWD}(T) = \text{NJTSSWD} \times \left[1 + \text{TNJTSSWD} \cdot \left(\frac{T}{\text{TNOM}} - 1\right)\right], \tag{12.94}$$

$$\text{NJTSSWGD}(T) = \text{NJTSSWGD} \times \left[1 + \text{TNJTSSWGD} \cdot \left(\frac{T}{\text{TNOM}} - 1\right)\right], \tag{12.95}$$

where NJTSD corresponds to the junction bottom, NJTSSWD corresponds to the sidewall, and NJTSSWGD corresponds to the gate sidewall.

12.8 SELF-HEATING EFFECT

Power density has been increasing with new technology nodes. The power is dissipated in the channel in the form of heat. As discussed and modeled earlier, the increase in temperature affects device parameters such as the mobility, threshold voltage, which results in a decrease in transistor current. The self-heating effect depends on the power dissipated and thermal conductivity of the materials used in the device. This is the reason, why this effect is more prominent in high-voltage/high-power devices—for example, laterally diffused MOS and insulated-gate bipolar transistor devices—as well as in silicon-on-insulator (SOI) MOSFETs owing to low thermal conductivity of the SOI substrate compared with bulk silicon. FinFETs also show a self-heating effect as the drive current is large and there is not much space to release this heat quickly from the device.

The self-heating effect is modeled using a thermal network [14] as shown in Figure 12.1 by invoking SHMOD = 1. The power dissipated in the transistor is fed into this thermal network and the voltage drop across the network results in an increase in temperature owing to the self-heating effect. This increase in temperature is added to the device temperature. SPICE optimizes the temperature increase and current decrease due to the self-heating effect till convergence is achieved. The increased number of iterations by SPICE cause a speed penalty.

The values of R_{th} and C_{th} scale with width and are obtained using following equations:

$$R_{th} = \frac{RTH0}{WTH0 + F_{pitch} \cdot NFIN}, \tag{12.96}$$

$$C_{th} = CTH0 \cdot (WTH0 + F_{pitch} \cdot NFIN), \tag{12.97}$$

where RTH0 and CTH0 are model parameters corresponding to normalized thermal resistance and thermal capacitance, respectively. WTH0 refers to the minimum width for calculation of the thermal resistance.

Figure 12.2 shows the simulation of drain current with drain voltage with and without the self-heating effect. When self-heating is turned on using SHMOD = 1,

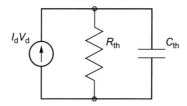

FIGURE 12.1

Thermal network for modeling the self-heating effect.

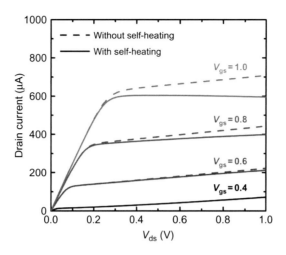

FIGURE 12.2

$I_d - V_d$ simulation with and without self-heating. The drain current decreases owing to the self-heating effect and may cause negative g_{ds}.

the drain current is reduced compared with the case of SHMOD = 0. If the thermal resistance (RTH0) is sufficiently large, it may even cause a decrease in current with increasing drain voltage, thereby causing negative g_{ds}. The benchmark test to check correct implementation of the self-heating effect was discussed in Chapter 10.

12.9 VALIDATION RANGE

The model is valid for temperatures ranging from −50 to 200 °C . For temperatures much lower than −50 °C, the device physics change considerably and modeling for such low temperatures is a separate, challenging issue.

12.10 MODEL VALIDATION ON MEASURED DATA

The temperature effects discussed in previous sections are validated on measured data of SOI FinFETs with $L = 60$ nm, TFIN = 17 nm, HFIN = 60 nm, and NFIN = 20. The temperature is swept from −50 to 200 °C in steps of 50 °C. Figure 12.3 shows the drain current in the linear region for different temperatures. The threshold voltage for different temperatures matches perfectly. At low gate voltage, mobility is strongly affected by temperature, while at high gate voltage, series resistances dominate the linear current characteristics. The model also matches the zero-temperature-coefficient, which is important in designing temperature-invariant circuits. Figure 12.4 shows the drain current in the saturation region for different temperatures.

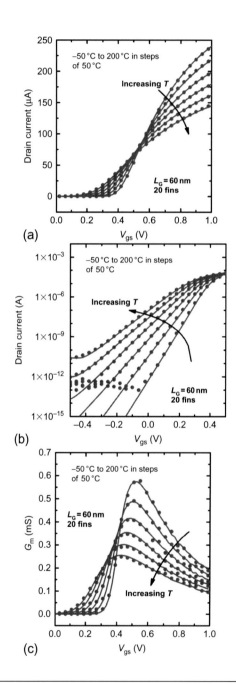

FIGURE 12.3

(a) Drain current on a linear axis, (b) drain current on a log axis, and (c) transconductance versus gate bias in the linear region of operation ($V_{ds} = 50\,\text{mV}$) for $T = -50\,°\text{C}$ to $200\,°\text{C}$ in steps of $50\,°\text{C}$. The experimental data are from an SOI FinFET with $L = 60\,\text{nm}$, TFIN $= 17\,\text{nm}$, HFIN $= 60\,\text{nm}$, and NFIN $= 20$.

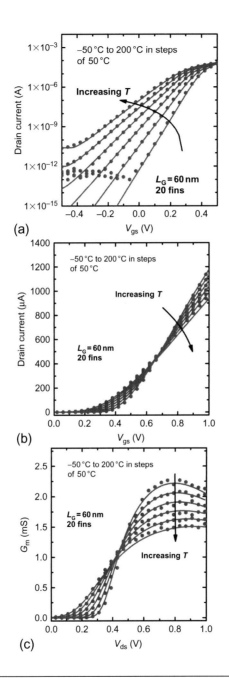

FIGURE 12.4

(a) Drain current on a linear axis, (b) drain current on a log axis, and (c) transconductance versus gate bias in the saturation region of operation ($V_{ds} = 1.0$ V) for $T = -50$ to 200 °C in steps of 50 °C. The experimental data are from an SOI FinFET with $L = 60$ nm, TFIN $= 17$ nm, HFIN $= 60$ nm, and NFIN $= 20$.

In saturation, the current is dominated by velocity saturation and the model shows an excellent match with experimental data.

REFERENCES

[1] C. Hu, Modern Semiconductor Devices for Integrated Circuits, Prentice Hall, New Jersey, 2010.

[2] H.Y. Fan, Temperature dependence of the energy gap in semiconductors, Phys. Rev. 82 (6) (1951) 900–905.

[3] B.V. Zeghbroeck, Principles of semiconductor devices, [Online] Available URL: http://ecee.colorado.edu/~bart/book/contents.htm.

[4] Y.P. Varshni, Temperature dependence of the energy gap in semiconductors, Physica 34 (1) (1967) 149–154.

[5] S.M. Sze, Physics of Semiconductor Devices, Wiley, New York, 1981.

[6] Y. Cheng, et al., Modeling temperature effects of quarter micrometer MOSFETs in BSIM3v3 for circuit simulation, Semicond. Sci. Technol. 12 (1997) 1349–1354.

[7] J.H. Huang, et al., BSIM3 Manual (Version 2.0), University of California, Berkeley, March 1994.

[8] Y. Cheng, et al., BSIM3 version 3.1 Users Manual, University of California, Berkeley, Memorandum No. UCB/ERL M97/2, 1997.

[9] M.S. Liang, J.Y. Choi, P.K. Ko, C. Hu, Inversion-layer capacitance and mobility of very thin gate-oxide MOSFETs, IEEE Trans. Electron Dev. ED-33 (1986) 409.

[10] C.L. Huang, G.S. Gildenblat, Measurements and modeling of the n-channel MOSFET inversion layer mobility and device characteristics in the temperature range 60-300K, IEEE Trans. Electron Devices ED-37 (1990) 1289–1300.

[11] Y. Cheng, C. Hu, MOSFET Modeling and BSIM3 User's Guide, Kluwer Academic Publishers, Norwell, MA, 1999.

[12] G. Massobrio, P. Antognetti, Semiconductor Device Modeling with SPICE, McGraw-Hill, Inc., New York, 1993.

[13] N. Arora, MOSFET Models for VLSI Circuit Simulation, Springer-Verlag, Vienna, New York, 1994.

[14] P. Su, S.K.H. Fung, S. Tang, F. Assaderaghi, C. Hu, BSIMPD: a partial-depletion SOI MOSFET model for deep-submicron CMOS designs, Proceedings of the IEEE Custom Integrated Circuits Conference, 2000, pp. 197–200.

Parameter list

CHAPTER OUTLINE

A.1 Model controllers 261
A.2 Instance parameters 263
A.3 Process parameters 264
A.4 Basic model parameters 266
A.5 Parameters for geometry-dependent parasitics 280
A.6 Parameters for temperature dependence and self-heating 281
A.7 Parameters for variability modeling 285

A.1 MODEL CONTROLLERS

Name	Unit	Default	Min	Max	Description
TYPE	–	NMOS	PMOS	NMOS	NMOS = 1, PMOS = 0
BULKMOD	–	0	0	1	Substrate model selector; 0 = multi-gate on SOI substrate, 1 = multi-gate on bulk substrate.
COREMOD	–	0	0	1	Simplified surface potential selector; 0 = turn off, 1 = turn on (lightly-doped or undoped)
GEOMOD	–	1	0	3	Structure selector; 0 = double gate, 1 = triple gate, 2 = quadruple gate, 3 = cylindrical gate
CGEOMOD	–	0	0	1	For CGEOMOD = 1 only, GEO1SW = 1 enables the parameters COVS, COVD, CGSP, and CGDP to be in F per fin, per gate-finger, per unit channel width

Continued

Name	Unit	Default	Min	Max	Description
RDSMOD	–	0	0	1	Bias-dependent, source/drain extension resistance model selector; 0 = internal bias dependent, 1 = external, 2 = internal
ASYMMOD	–	0	0	1	Asymmetric I-V model selector; 0 = turn off, reverse mode parameters ignored, 1 = turn on
IGCMOD	–	0	0	1	Model selector for Igc, Igs, and Igd; 1 = turn on, 0 = turn off
IGBMOD	–	0	0	1	Model selector for Igb; 1 = turn on, 0 = turn off
GIDLMOD	–	0	0	1	GIDL/GISL current switcher; 1 = turn on, 0 = turn off
IIMOD	–	0	0	2	Impact ionization model switch; 0 = OFF, 1 = BSIM4 based, 2 = BSIMSOI based
NQSMOD	–	0	0	1	NQS gate resistor and *gi* node switcher; 1 = turn on, 0 = turn off
SHMOD	–	0	0	1	Self-heating and *T* node switcher; 1 = turn on, 0 = turn off
RGATEMOD	–	0	0	1	Gate electrode resistor and *ge* node switcher; 1 = turn on, 0 = turn off
RGEOMOD	–	0	0	1	Bias independent parasitic resistance model selector
CGEOMOD	–	0	0	2	Parasitic capacitance model selector
CAPMOD	–	0	0	1	Accumulation region capacitance model selector; 0 = no accumulation capacitance, 1 = accumulation capacitance included
TEMPMOD	–	0	0	1	Temperature dependence model selector
TNOIMOD	–	0	0	2	Thermal noise model selector; 0 = charge-based, 1 = holistic, 2 = correlated noise model

A.2 **INSTANCE PARAMETERS**

Name	Unit	Default	Min	Max	Description
$L^{(m)}$	m	30n	1n	–	Designed gate length
$D^{(m)}$	m	40n	1n	–	Diameter of cylinder (for GEOMOD = 3)
$TFIN^{(m)}$	m	15n	1n	–	Body (fin) thickness
$FPITCH^{(m)}$	m	80n	TFIN	–	Fin pitch
NF	–	1	1	–	Number of fingers
$NFIN^{(m)}$	–	1	>0	–	Number of fins per finger
$NGCON^{(m)}$	–	1	1	2	Number of gate contacts
$ASEO^{(m)}$	m^2	0	0	–	Source to substrate overlap area through oxide (all fingers)
$ADEO^{(m)}$	m^2	0	0	–	Drain to substrate overlap area through oxide (all fingers)
$PSEO^{(m)}$	m	0	0	–	Perimeter of source to substrate overlap region through oxide (all fingers)
$PDEO^{(m)}$	m	0	0	–	Perimeter of drain to substrate overlap region through oxide (all fingers)
$ASEJ^{(m)}$	m^2	0	0	–	Source junction area (all fingers; for bulk MuGFETs, BULKMOD = 1)
$ADEJ^{(m)}$	m^2	0	0	–	Drain junction area (all fingers; for bulk MuGFETs, BULKMOD = 1)
$PSEJ^{(m)}$	m	0	0	–	Source junction perimeter (all fingers; for bulk MuGFETs, BULKMOD = 1)
$PDEJ^{(m)}$	m	0	0	–	Drain junction perimeter (all fingers; for bulk MuGFETs, BULKMOD = 1)
$COVS^{(m)}$	F or F/m see CGEO1SW	0	0	–	Constant gate to source overlap capacitance (for CGEOMOD = 1)
$COVD^{(m)}$	F or F/m see CGEO1SW	CVOS	0	–	Constant gate to drain overlap capacitance (for CGEOMOD = 1)

Continued

Name	Unit	Default	Min	Max	Description
CGSP$^{(m)}$	F or F/m see CGEO1SW	0	0	–	Constant gate to source fringe capacitance (for CGEOMOD = 1)
CGDP$^{(m)}$	F or F/m see CGEO1SW	0	0	–	Constant gate to drain fringe capacitance (for CGEOMOD = 1)
CDSP$^{(m)}$	F	0	0	–	Constant drain to source fringe capacitance
NRS$^{(m)}$	–	0	0	–	Number of source diffusion squares (for RGEOMOD = 0)
NRD$^{(m)}$	–	0	0	–	Number of drain diffusion squares (for RGEOMOD = 0)
LRSD$^{(m)}$	m	L	0	–	Length of the source/drain

Note: Instance parameters with superscript $^{(m)}$ are also model parameters.

A.3 PROCESS PARAMETERS

Name	Unit	Default	Min	Max	Description
XL	m	0	–	–	L offset for channel length due to mask/etch effect
LINT	m	0.0	–	–	Length reduction parameter (dopant diffusion effect)
LL	m$^{(LLN+1)}$	0.0	–	–	Length reduction parameter (dopant diffusion effect)
LLN	–	1.0	–	–	Length reduction parameter (dopant diffusion effect)
DLC	m	0.0	–	–	Length reduction parameter for CV (dopant diffusion effect)
DLCACC	m	0.0	–	–	Length reduction parameter for CV in accumulation region (BULKMOD = 1, CAPMOD = 1)
LLC	m$^{(LLN+1)}$	0.0	–	–	Length reduction parameter for CV (dopant diffusion effect)
DLBIN	m	0.0	–	–	Length reduction parameter for binning
EOT	m	1.0n	0.1n	–	SiO$_2$ equivalent gate dielectric thickness (including inversion layer thickness)
TOXP	m	1.2n	0.1n	–	Physical oxide thickness

Name	Unit	Default	Min	Max	Description
EOTBOX	m	140n	1n	–	SiO$_2$ equivalent buried oxide thickness (including substrate depletion)
HFIN	m	30n	1n	–	Fin height
FECH	–	1.0	0	–	End-channel factor, for different orientation/shape (mobility difference between the side channel and the top channel is handled by this parameter)
DELTAW	m	0.0	–	–	Reduction of effective width due to shape of fin
FECHCV	–	1.0	0	–	CV end-channel factor, for different orientation/shape
DELTAWCV	m	0.0	–	–	CV reduction of effective width due to shape of fin
NBODY	m^{-3}	1e22	–	–	Channel (body) doping concentration
NBODYN1	–	0	−0.08	–	NFIN dependence of NBODY
NBODYN2	–	1e5	1e−5	–	NFIN dependence of NBODY
NSD	m^{-3}	2e26	2e25	1e27	S/D doping concentration
PHIG	eV	4.61	0	–	Gate workfunction
PHIGL	eV/m	0	–	–	Length dependence of gate workfunction
PHIGN1	–	0	−0.08	–	NFIN dependence of PHIG
PHIGN2	–	1e5	1e−5	–	NFIN dependence of PHIG
EPSROX	–	3.9	1	–	Relative dielectric constant of the gate insulator
EPSRSUB	–	11.9	1	–	Relative dielectric constant of the channel material
EASUB	eV	4.05	0	–	Electron affinity of the substrate material
NI0SUB	m^{-3}	1.1e16	–	–	Intrinsic carrier concentration of channel at 300.15K
BG0SUB	eV	1.12	–	–	Band gap of the channel material at 300.15K
NC0SUB	m^{-3}	2.86e25	–	–	Conduction band density of states at 300.15K
NGATE	m^{-3}	0	–	–	Parameter for Poly Gate doping. Set NGATE = 0 for metal gates
IMIN	A/m^2	1e−15	–	–	Parameter for voltage clamping for inversion region calc. in accumulation

A.4 BASIC MODEL PARAMETERS

Name	Unit	Default	Min	Max	Description
CIT	F/m^2	0.0	–	–	Parameter for interface trap
CDSC	F/m^2	7e−3	0.0	–	Coupling capacitance between S/D and channel
CDSCN1	–	0	−0.08	–	NFIN dependence of CDSC
CDSCN2	–	1e5	1e−5	–	NFIN dependence of CDSC
CDSCD	F/m^2	7e−3	0.0	–	Drain-bias sensitivity of CDSC
CDSCDN1	–	0	−0.08	–	NFIN dependence of CDSCD
CDSCDN2	–	1e5	1e−5	–	NFIN dependence of CDSCD
CDSCDR	F/m^2	CDSCD	0.0	–	Reverse-mode drain-bias sensitivity
CDSCDRN1	–	CDSCDN1	−0.08	–	NFIN dependence of CDSCDR
CDSCDRN2	–	CDSCDN2	1e−5	–	NFIN dependence of CDSCDR
DVT0	–	0.0	0.0	–	SCE coefficient
DVT1	–	0.60	>0	–	SCE exponent coefficient
DVT1SS	–	DVT1	>0	–	Subthreshold swing exponent coefficient
PHIN	V	0.05	–	–	Nonuniform vertical doping effect on surface potential
ETA0	–	0.60	0.0	–	DIBL coefficient
ETA0N1	–	0	−0.08	–	NFIN dependence of ETA0
ETA0N2	–	0	1e−5	–	NFIN dependence of ETA0
DSUB	–	1.06	>0	–	DIBL exponent coefficient
DVTP0	–	0	–	–	Coefficient for drain-induced V_{th} shift (DITS)
DVTP1	–	0	–	–	DITS exponent coefficient
K1RSCE	V$^{1/2}$	0.0	–	–	Prefactor for reverse short channel effect
LPE0	m	5e−9	−L_{eff}	–	Equivalent length of pocket region at zero bias
K0	V	–	0.0	–	Lateral NUD parameter
K0SI	–	1.0	>0	–	Correction factor for strong inversion/g_m
K1SI	–	K0SI	>0	–	Correction factor for strong inversion, used in M_{ob}

Name	Unit	Default	Min	Max	Description
DVTSHIFT	V	0.0	–	–	Additional V_{th} shift handle
PHIBE	V	0.7	0.2	1.2	Body-effect voltage parameter
K1	$V^{1/2}$	0.0	0.0	–	Body-effect coefficient for subthreshold region
K1SAT	$V^{-1/2}$	0.0	–	–	Body-effect coefficient for saturation region
QMFACTOR	–	0.0	–	–	Prefactor for QM V_{th} shift correction
QMTCENIV	–	0.0	–	–	Prefactor/switch for QM effective width correction for IV
QMTCENCV	–	0.0	–	–	Prefactor/switch for QM effective width and oxide thickness correction for CV
QMTCENCVA	–	0.0	–	–	Prefactor/switch for QM effective width and oxide thickness correction for accumulation region CV
ETAQM	–	0.54	–	–	Body-charge coefficient for QM charge centroid
QM0	V	1e–3	>0	–	Normalization parameter for QM charge centroid (inversion)
PQM	–	0.66	–	–	Fitting parameter for QM charge centroid (inversion)
QM0ACC	V	1e–3	>0	–	Normalization parameter for QM charge centroid (accumulation)
PQMACC	–	0.66	–	–	Fitting parameter for QM charge centroid (accumulation)
VSAT	m/s	85,000	–	–	Saturation velocity for the saturation region
VSATN1	–	0	–0.08	–	NFIN dependence of VSAT
VSATN2	–	1e5	1e–5	–	NFIN dependence of VSAT
VSAT1	m/s	VSAT	–	–	Saturation velocity for the linear region in forward mode
VSAT1N1	–	0	–0.08	–	NFIN dependence of VSAT1
VSAT1N2	–	1e5	1e–5	–	NFIN dependence of VSAT1

Continued

Name	Unit	Default	Min	Max	Description
VSAT1R	m/s	VSAT1	–	–	Saturation velocity for the linear region in reverse mode
VSAT1RN1	–	VSAT1N1	−0.08	–	NFIN dependence of VSAT1R
VSAT1RN2	–	VSAT1N2	1e−5	–	NFIN dependence of VSAT1R
DELTAVSAT	–	1.0	0.01	–	Velocity saturation parameter in the linear region
PSAT	–	2.0	2.0	–	Exponent for field for velocity saturation
KSATIV	–	1.0	–	–	Parameter for long channel V_{dsat}
VSATCV	m/s	VSAT	–	–	Saturation velocity for the capacitance model
DELTAVSATCV	–	DELTAVSAT	0.01	–	Velocity saturation parameter in the linear region for the capacitance model
PSATCV	–	PSAT	2.0	–	Exponent for field for velocity saturation for the capacitance model
MEXP	–	4	2	–	Smoothing function factor for V_{dsat}
MEXPR	–	MEXP	2	–	Reverse-mode smoothing function factor for V_{dsat}
PTWG	V^{-2}	0.0	–	–	Correction factor for velocity saturation in forward mode
PTWGR	V^{-2}	PTWG	–	–	Correction factor for velocity saturation in reverse mode
A1	V^{-2}	0.0	–	–	Non-saturation effect parameter in strong inversion region
A2	V^{-1}	0.0	–	–	Non-saturation effect parameter in moderate inversion region
U0	m^2/Vs	3e−2	–	–	Low field mobility
U0N1	–	0	−0.08	–	NFIN dependence of U0
U0N2	–	1e5	1e−5	–	NFIN dependence of U0

Name	Unit	Default	Min	Max	Description
CHARGEWF	–	0	−1	1	Average channel charge weighting (sampling) factor, +1: source-side, 0 : middle, −1 : drain-side
ETAMOB	–	2.0	–	–	Effective field parameter
UP	μm^{LPA}	0.0	–	–	Mobility L coefficient
LPA	–	1.0	–	–	Mobility L power coefficient
UA	$(cm/MV)^{EU}$	0.3	>0.0	–	Phonon/surface roughness scattering parameter
UC	$(1.0e{-}6{*}cm/MV^2)^{EU}$	0.0	–	–	Body effect coefficient for mobility (BULKMOD = 1)
EU	cm/MV	2.5	>0.0	–	Phonon/surface roughness scattering parameter
UD	cm/MV	0.0	>0.0	–	Columbic scattering parameter
UCS	–	1.0	>0.0	–	Columbic scattering parameter
PCLM	–	0.013	>0.0	–	Channel length modulation (CLM) parameter
PCLMG	–	0	–	–	Gate bias dependent parameter for channel length modulation (CLM)
RDSWMIN	$\Omega\,\mu_m^{WR}$	0.0	0.0	–	RDSMOD = 0 S/D extension resistance per unit width at high V_{gs}
RDSW	$\Omega\,\mu_m^{WR}$	100	0.0	–	RDSMOD = 0 zero bias S/D extension resistance per unit width
RSWMIN	$\Omega\,\mu_m^{WR}$	0.0	0.0	–	RDSMOD = 1 source extension resistance per unit width at high V_{gs}
RSW	$\Omega\,\mu_m^{WR}$	50	0.0	–	RDSMOD = 1 zero bias source extension resistance per unit width
RDWMIN	$\Omega\,\mu_m^{WR}$	0.0	0.0	–	RDSMOD = 1 drain extension resistance per unit width at high V_{gs}

Continued

Name	Unit	Default	Min	Max	Description
RDW	$\Omega\,\mu_m^{WR}$	50	0.0	–	RDSMOD = 1 zero bias drain extension resistance per unit width
RSDR	V^{-PRSDR}	0.0	0.0	–	RDSMOD = 1 source side drift resistance parameter in forward mode
RSDRR	V^{-PRSDR}	RSDR	0.0	–	RDSMOD = 1 source side drift resistance parameter in reverse mode
RDDR	V^{-PRDDR}	RSDR	0.0	–	RDSMOD = 1 drain side drift resistance parameter in forward mode
RSDRR	V^{-PRDDR}	RDDR	0.0	–	RDSMOD = 1 drain side drift resistance parameter in reverse mode
PRWGS	V^{-1}	0.0	0.0	–	Source side quasi-saturation parameter
PRWGD	V^{-1}	PRWGS	0.0	–	Drain side quasi-saturation parameter
PRSDR	–	1.0	0.0	–	RDSMOD = 1 drain side drift resistance parameter in forward mode
PRDDR	–	PRSDR	0.0	–	RDSMOD = 1 drain side drift resistance parameter in reverse mode
WR	–	1.0	–	–	W dependence parameter of S/D extension resistance
RGEXT	Ω	0.0	0.0	–	Effective gate electrode external resistance (experimental)
RGFIN	Ω	1.0e−3	1.0e−3	–	Effective gate electrode resistance per fin per finger
RSHS	Ω	0.0	0.0	–	Source-side sheet resistance
RSHD	Ω	RSHS	0.0	–	Drain-side sheet resistance
PDIBL1	–	1.30	0.0	–	Parameter for DIBL effect on rout in forward mode
PDIBL1R	–	PDIBL1	0.0	–	Parameter for DIBL effect on rout in reverse mode
PDIBL2	–	2e−4	0.0	–	Parameter for DIBL effect on rout
DROUT	–	1.06	>0.0	–	L dependence of DIBL effect on rout
PVAG	–	1.0	–	–	V_{gs} dependence on early voltage

Name	Unit	Default	Min	Max	Description
TOXREF	m	1.2nm	>0.0	–	Nominal gate oxide thickness for gate tunneling current
TOXG	m	TOXP	>0.0	–	Oxide thickness for gate current model
NTOX	–	1.0	–	–	Exponent for gate oxide ratio
AIGBINV	$(Fs^2/g)^{0.5}m^{-1}$	1.11e−2	–	–	Parameter for Igb in inversion
BIGBINV	$(Fs^2/g)^{0.5}m^{-1}V^{-1}$	9.49e−4	–	–	Parameter for Igb in inversion
CIGBINV	V^{-1}	6.00e−3	–	–	Parameter for Igb in inversion
EIGBINV	V	1.1	–	–	Parameter for Igb in inversion
NIGBINV	–	3.0	>0.0	–	Parameter for Igb in inversion
AIGBACC	$(Fs^2/g)^{0.5}m^{-1}$	1.36e−2	–	–	Parameter for Igb in accumulation
BIGBACC	$(Fs^2/g)^{0.5}m^{-1}V^{-1}$	1.71e−3	–	–	Parameter for Igb in accumulation
CIGBACC	V^{-1}	7.5e−2	–	–	Parameter for Igb in accumulation
NIGBACC	–	1.0	>0.0	–	Parameter for Igb in accumulation
AIGC	$(Fs^2/g)^{0.5}m^{-1}$	1.36e−2	–	–	Parameter for Igc in inversion
BIGC	$(Fs^2/g)^{0.5}m^{-1}V^{-1}$	1.71e−3	–	–	Parameter for Igc in inversion
CIGC	V^{-1}	0.075	–	–	Parameter for Igc in inversion
PIGCD	–	1.0	>0.0	–	V_{ds} dependence of Igcs and Igcd
DLCIGS	m	0.0	–	–	Delta L for Igs model
AIGS	$(Fs^2/g)^{0.5}m^{-1}$	1.36e−2	–	–	Parameter for Igs in inversion
BIGS	$(Fs^2/g)^{0.5}m^{-1}V^{-1}$	1.71e−3	–	–	Parameter for Igs in inversion
CIGS	V^{-1}	0.075	–	–	Parameter for Igs in inversion
DLCIGD	m	DLCIGS	–	–	Delta L for Igd model

Continued

Name	Unit	Default	Min	Max	Description
AIGD	$(Fs^2/g)^{0.5}m^{-1}$	AIGS	–	–	Parameter for Igd in inversion
BIGD	$(Fs^2/g)^{0.5}m^{-1}V^{-1}$	BIGS	–	–	Parameter for Igd in inversion
CIGD	V^{-1}	CIGS	–	–	Parameter for Igd in inversion
POXEDGE	–	1	>0.0	–	Factor for the gate edge Tox
AGIDL	Ω^{-1}	6.055e−12	–	–	Pre-exponential coefficient for GIDL
BGIDL	V/m	0.3e9	–	–	Exponential coefficient for GIDL
CGIDL	V^3	0.2	–	–	Parameter for body bias effect of GIDL
EGIDL	V	0.2	–	–	Band bending parameter for GIDL
PGIDL	–	1.0	–	–	Exponent of electric field for GIDL
AGISL	Ω^{-1}	AIGDL	–	–	Pre-exponential coefficient for GISL
BGISL	V/m	BGIDL	–	–	Exponential coefficient for GISL
CGISL	V^3	0.2	–	–	Parameter for body bias effect of GISL
EGISL	V	EGIDL	–	–	Band bending parameter for GISL
PGISL	–	1.0	–	–	Exponent of electric field for GISL
ALPHA0	mV^{-1}	0.0	–	–	First parameter of I_{ii} (IIMOD = 1)
ALPHA1	V^{-1}	0.0	–	–	L scaling parameter of I_{ii} (IIMOD = 1)
ALPHAII0	mV^{-1}	0.0	–	–	First parameter of I_{ii} (IIMOD = 2)
ALPHAII1	V^{-1}	0.0	–	–	L scaling parameter of I_{ii} (IIMOD = 2)
BETA0	V^{-1}	0.0	–	–	V_{ds} dependent parameter of I_{ii} (IIMOD = 1)
BETAII0	V^{-1}	0.0	–	–	V_{ds} dependent parameter of I_{ii} (IIMOD = 2)
BETAII1	–	0.0	–	–	V_{ds} dependent parameter of I_{ii} (IIMOD = 2)
BETAII2	V	0.1	–	–	V_{ds} dependent parameter of I_{ii} (IIMOD = 2)

Name	Unit	Default	Min	Max	Description
ESATII	V/m	1.0e7	–	–	Saturation channel E-field for I_{ii} (IIMOD = 2)
LII	Vm	0.5e−9	–	–	Channel length dependent parameter of I_{ii} (IIMOD = 2)
SII0	V^{-1}	0.5	–	–	V_{gs} dependent parameter of I_{ii} (IIMOD = 2)
SII1	–	0.1	–	–	V_{gs} dependent parameter of I_{ii} (IIMOD = 2)
SII2	V	0.0	–	–	V_{gs} dependent parameter of I_{ii} (IIMOD = 2)
SIID	V	0.0	–	–	V_{ds} dependent parameter of I_{ii} (IIMOD = 2)
EOTACC	m	EOT	0.1n	–	SiO_2 equivalent gate dielectric thickness for accumulation region
DELVFBACC	V	0.0	–	–	Additional V_{fb} shift required for accumulation region
PCLMCV	–	0.013	>0.0	–	Channel length modulation (CLM) parameter for the capacitance model
CFS	F/m	2.5e−11	0.0	–	Source-side outer fringe cap (for CGEOMOD = 0)
CFD	F/m	CFS	0.0	–	Drain-side outer fringe cap (for CGEOMOD = 0)
CGSO	F/m	Calculated	0.0	–	Non LDD region source-gate overlap capacitance per unit channel width (for CGEOMOD =0,2)
CGDO	F/m	Calculated	0.0	–	Non LDD region drain-gate overlap capacitance per unit channel width (for CGEOMOD = 0,2)
CGSL	F/m	0	0.0	–	Overlap capacitance between gate and lightly-doped source region (for CGEOMOD = 0,2)
CGDL	F/m	CGSL	0.0	–	Overlap capacitance between gate and lightly-doped drain region (for CGEOMOD = 0,2)
CKAPPAS	V	0.6	0.02	–	Coefficient of bias-dependent overlap capacitance for the source side (for CGEOMOD = 0,2)

Continued

Name	Unit	Default	Min	Max	Description
CKAPPAD	V	CKAPPAS	0.02	–	Coefficient of bias-dependent overlap capacitance for the drain side (for CGEOMOD = 0,2)
CGBO	F/m	0	0.0	–	Gate-substrate overlap capacitance per unit channel length per finger per gate contact
CGBN	F/m	0	0.0	–	Gate-substrate overlap capacitance per unit channel length per finger per fin
CSDESW	F/m	0	0.0	–	Source/drain sidewall fringing capacitance per unit length
CJS	F/m^2	0.0005	0.0	–	Unit area source-side junction capacitance at zero bias
CJD	F/m^2	CJS	0.0	–	Unit area drain-side junction capacitance at zero bias
CJSWS	F/m	5.0e−10	0.0	–	Unit length sidewall junction capacitance at zero bias (source-side)
CJSWD	F/m	CJSWS	0.0	–	Unit length sidewall junction capacitance at zero bias (drain-side)
CJSWGS	F/m	0.0	0.0	–	Unit length gate sidewall junction capacitance at zero bias (source-side)
CJSWGD	F/m	CJSWGS	0.0	–	Unit length gate sidewall junction capacitance at zero bias (drain-side)
PBS	V	1.0	0.01	–	Bottom junction built-in potential (source-side)
PBD	V	PBS	0.01	–	Bottom junction built-in potential (drain-side)
PBSWS	V	1.0	0.01	–	Isolation-edge sidewall junction built-in potential (source-side)
PBSWD	V	PBSWS	0.01	–	Isolation-edge sidewall junction built-in potential (drain-side)
PBSWGS	V	PBSWS	0.01	–	Gate-edge sidewall junction built-in potential (source-side)

Name	Unit	Default	Min	Max	Description
PBSWGD	V	PBSWGS	0.01	–	Gate-edge sidewall junction built-in potential (drain-side)
MJS	–	0.5	–	–	Source bottom junction capacitance grading coefficient
MJD	–	MJS	–	–	Drain bottom junction capacitance grading coefficient
MJSWS	–	0.33	–	–	Isolation-edge sidewall junction capacitance grading coefficient (source-side)
MJSWD	–	MJSWS	–	–	Isolation-edge sidewall junction capacitance grading coefficient (drain-side)
MJSWGS	–	MJSWS	–	–	Gate-edge sidewall junction capacitance grading coefficient (source-side)
MJSWGD	–	MJSWGS	–	–	Gate-edge sidewall junction capacitance grading coefficient (drain-side)
SJS	–	0.0	0.0	–	Constant for source-side two-step second junction capacitance
SJD	–	SJS	0.0	–	Constant for drain-side two-step second junction capacitance
SJSWS	–	0.0	0.0	–	Constant for sidewall two-step second junction capacitance (source-side)
SJSWD	–	SJSWS	0.0	–	Constant for sidewall two-step second junction capacitance (drain-side)
SJSWGS	–	0.0	0.0	–	Constant for gate sidewall two-step second junction capacitance (source-side)
SJSWGD	–	SJSWGS	0.0	–	Constant for gate sidewall two-step second junction capacitance (drain-side)
MJS2	–	0.125	–	–	Source bottom two-step second junction capacitance grading coefficient

Continued

Name	Unit	Default	Min	Max	Description
MJD2	–	MJS2	–	–	Drain bottom two-step second junction capacitance grading coefficient
MJSWS2	–	0.083	–	–	Isolation-edge sidewall two-step second junction capacitance grading coefficient (source-side)
MJSWD2	–	MJSWS2	–	–	Isolation-edge sidewall two-step second junction capacitance grading coefficient (drain-side)
MJSWGS2	–	MJSWS2	–	–	Gate-edge sidewall two-step second junction capacitance grading coefficient (source-side)
MJSWGD2	–	MJSWGS2	–	–	Gate-edge sidewall two-step second junction capacitance grading coefficient (drain-side)
JSS	A/m^2	1.0e−4	0.0	–	Bottom source junction reverse saturation current density
JSD	A/m^2	JSS	0.0	–	Bottom drain junction reverse saturation current density
JSWS	A/m	0	0.0	–	Unit length reverse saturation current for isolation-edge source sidewall junction
JSWD	A/m	JSWS	0.0	–	Unit length reverse saturation current for isolation-edge drain sidewall junction
JSWGS	A/m	0	0.0	–	Unit length reverse saturation current for gate-edge source sidewall junction
JSWGD	A/m	JSWGS	0.0	–	Unit length reverse saturation current for gate-edge drain sidewall junction
JTSS	A/m^2	0	0.0	–	Bottom source junction trap-assisted saturation current density

Name	Unit	Default	Min	Max	Description
JTSD	A/m^2	JTSS	0.0	–	Bottom drain junction trap-assisted saturation current density
JTSSWS	A/m	0	0.0	–	Unit length trap-assisted saturation current for isolation-edge source sidewall junction
JTSSWD	A/m	JTSSWS	0.0	–	Unit length trap-assisted saturation current for isolation-edge drain sidewall junction
JTSSWGS	A/m	0	0.0	–	Unit length trap-assisted saturation current for gate-edge source sidewall junction
JTSSWGD	A/m	JTSSWGS	0.0	–	Unit length trap-assisted saturation current for gate-edge drain sidewall junction
JTWEFF	m	0	0.0	–	Trap assisted tunneling current width dependence
NJS	–	1.0	0.0	–	Source junction emission coefficient
NJD	–	NJS	0.0	–	Drain junction emission coefficient
NJTS	–	20	0.0	–	Non-ideality factor for JTSS
NJTSD	–	NJTS	0.0	–	Non-ideality factor for JTSD
NJTSSW	–	20	0.0	–	Non-ideality factor for JTSSWS
NJTSSWD	–	NJTSSW	0.0	–	Non-ideality factor for JTSSWD
NJTSSWG	–	20	0.0	–	Non-ideality factor for JTSSWGS
NJTSSWGD	–	NJTSSWG	0.0	–	Non-ideality factor for JTSSWGD
VTSS	V	10	0.0	–	Bottom source junction trap-assisted current voltage dependent parameter
VTSD	V	VTSS	0.0	–	Bottom drain junction trap-assisted current voltage dependent parameter

Continued

Name	Unit	Default	Min	Max	Description
VTSSWS	V	10	0.0	–	Unit length trap-assisted current voltage dependent parameter for sidewall source junction
VTSSWD	V	VTSSWS	0.0	–	Unit length trap-assisted current voltage dependent parameter for sidewall drain junction
VTSSWGS	V	10	0.0	–	Unit length trap-assisted current voltage dependent parameter for gate-edge sidewall source junction
VTSSWGD	V	VTSSWGS	0.0	–	Unit length trap-assisted current voltage dependent parameter for gate-edge sidewall drain junction
IJTHSFWD	A	0.1	$10 I_{sbs}$	–	Forward source diode breakdown limiting current
IJTHDFWD	A	IJTHSFWD	$10 I_{sbd}$	–	Forward drain diode breakdown limiting current
IJTHSREV	A	0.1	$10 I_{sbs}$	–	Reverse source diode breakdown limiting current
IJTHDREV	A	IJTHSREV	$10 I_{sbd}$	–	Reverse drain diode breakdown limiting current
BVS	V	10.0	–	–	Source diode breakdown voltage
BVD	V	BVS	–	–	Drain diode breakdown voltage
XJBVS	–	1.0	–	–	Fitting parameter for source diode breakdown current
XJBVD	–	XJBVS	–	–	Fitting parameter for source diode breakdown current
LINTIGEN	m	0.0	–	$L_{eff}/2$	L_{int} offset for R/G current
NTGEN	–	1.0	>0.0	–	Parameter for R/G current (experimental)
AIGEN	$m^{-3}V^{-1}$	0.0	–	–	Parameter for R/G current (experimental)
BIGEN	$m^{-3}V^{-3}$	0.0	–	–	Parameter for R/G current (experimental)

Name	Unit	Default	Min	Max	Description
XRCRG1	–	12.0	0.0 or $\geq 10^{-3}$	–	Parameter for non quasi-static gate resistance (NQSMOD = 1 and NQSMOD = 2)
XRCRG2	–	1.0	–	–	Parameter for non quasi-static gate resistance (NQSMOD = 1 and NQSMOD = 2)
NSEG	–	5	4	10	Number of channel segments for NQSMOD = 3
EF	–	1.0	>0.0	2.0	Flicker noise frequency exponent
LINTNOI	m	0.0	–	$L_{eff}/2$	L_{int} offset for flicker noise calculation
EM	V/m	4.1e7	–	–	Flicker noise parameter
NOIA	$eV^{-1}s^{1-EF}m^{-3}$	6.250e39	–	–	Flicker noise parameter
NOIB	$eV^{-1}s^{1-EF}m^{-1}$	3.125e24	–	–	Flicker noise parameter
NOIC	$eV^{-1}s^{1-EF}m$	8.750e7	–	–	Flicker noise parameter
NTNOI	–	1.0	0.0	–	Thermal noise parameter
RNOIA	–	0.577	–	–	Thermal noise parameter
RNOIB	–	0.37	–	–	Thermal noise parameter
TNOIA	m^{-1}	1.5	0.0	–	Thermal noise parameter
TNOIB	m^{-1}	3.5	0.0	–	Thermal noise parameter
NVTM	V	nkT/q	–	–	If provided NVTM will override nkT/q calculated in the model
THETASCE	–	Θ_{SCE}	–	–	If provided THETASCE will override Θ_{SCE} calculated in the model
THETASW	–	Θ_{SW}	–	–	If provided THETASW will override Θ_{SW} calculated in the model
THETADIBL	–	Θ_{DIBL}	–	–	If provided THETADIBL will override Θ_{DIBL} calculated in the model

A.5 PARAMETERS FOR GEOMETRY-DEPENDENT PARASITICS

The parameters listed in this section are for RGEOMOD = 1 and CGEOMOD = 2.

Name	Unit	Default	Min	Max	Description
HEPI	m	10n	–	–	Height of the raised source/drain on top of the fin
TSILI	m	10n	–	–	Thickness of the silicide on top of the raised source/drain
RHOC	$\Omega\,m^2$	1p	10^{-18}	10^{-9}	Contact resistivity at the silicon/silicide interface
RHORSD	$\Omega\,m$	Calculated	0	–	Average resistivity of silicon in the raised source/drain region
RHOEXT	$\Omega\,m$	RHORSD	0	–	Average resistivity of silicon in the fin extension region
CRATIO	–	0.5	0	1	Ratio of the corner area filled with silicon to the total corner area
DELTAPRSD	m	0.0	-FPITCH	–	Change in silicon/silicide interface length due to non-rectangular epi
SDTERM	–	0	0	1	Indicator of whether the source/drain are terminated with silicide
LSP	m	0.2(L+XL)	0	–	Thickness of the gate sidewall spacer
LDG	m	5n	0	–	Lateral diffusion gradient in the fin extension region
EPSRSP	–	3.9	1	–	Relative dielectric constant of the gate sidewall spacer material
TGATE	m	30n	0	–	Gate height on top of the hard mask
TMASK	m	30n	0	–	Height of the hard mask on top of the fin
ASILIEND	m^2	0	0	–	Extra silicide cross sectional area at the two ends of the FinFET
ARSDEND	m^2	0	0	–	Extra raised source/drain cross sectional area at the two ends of the FinFET

Name	Unit	Default	Min	Max	Description
PRSDEND	m	0	0	–	Extra silicon/silicide interface perimeter at the two ends of the FinFET
NSDE	m^{-3}	2×10^{25}	10^{25}	10^{26}	Active doping concentration at the channel edge
RGEOA	–	1.0	–	–	Fitting parameter for RGEOMOD = 1
RGEOB	m^{-1}	0	–	–	Fitting parameter for RGEOMOD = 1
RGEOC	m^{-1}	0	–	–	Fitting parameter for RGEOMOD = 1
RGEOD	m^{-1}	0	–	–	Fitting parameter for RGEOMOD = 1
RGEOE	m^{-1}	0	–	–	Fitting parameter for RGEOMOD = 1
CGEOA	–	1.0	–	–	Fitting parameter for CGEOMOD = 2
CGEOB	m^{-1}	0	–	–	Fitting parameter for CGEOMOD = 2
CGEOC	m^{-1}	0	–	–	Fitting parameter for CGEOMOD = 2
CGEOD	m^{-1}	0	–	–	Fitting parameter for CGEOMOD = 2
CGEOE	–	1.0	–	–	Fitting parameter for CGEOMOD = 2

A.6 PARAMETERS FOR TEMPERATURE DEPENDENCE AND SELF-HEATING

Name	Unit	Default	Min	Max	Description
TNOM	C	27	−273.15	–	Temperature at which the model is extracted (in Celsius)
TBGASUB	eV/K	7.02e−4	–	–	Bandgap temperature coefficient
TBGBSUB	K	1108.0	–	–	Bandgap temperature coefficient
KT1[b]	V	0.0	–	–	V_{th} temperature coefficient
KT1L	V m	0.0	–	–	V_{th} temperature coefficient
TSS	1/K	0.0	–	–	Subthreshold swing temperature coefficient

Continued

Name	Unit	Default	Min	Max	Description
TETA0	1/K	0.0	–	–	Temperature dependence of DIBL coefficient
TETA0R	1/K	0.0	–	–	Temperature dependence of reverse-mode DIBL coefficient
UTE	–	0.0	–	–	Mobility temperature coefficient
UTL	–	−1.5e−3	–	–	Mobility temperature coefficient
EMOBT	–	0.0	–	–	Temperature coefficient of ETAMOB
UA1	–	1.032e−3	–	–	Mobility temperature coefficient for UA
UC1	–	0.056e−9	–	–	Mobility temperature coefficient for UC
UD1	–	0.0	–	–	Mobility temperature coefficient
UCSTE	–	−4.775e−3	–	–	Mobility temperature coefficient
AT	1/K	−0.00156	–	–	Saturation velocity temperature coefficient
ATCV	1/K	AT	–	–	Saturation velocity temperature coefficient for $C - V$
A11	V^{-2}/K	0.0	–	–	Temperature dependence of non-saturation effect parameter for strong inversion region
A21	V^{-1}/K	0.0	–	–	Temperature dependence of non-saturation effect parameter for moderate inversion region
K01	V/K	0.0	–	–	Temperature dependence of K0
K0SI1	1/K	0.0	–	–	Temperature dependence of K0SI
K11	$V^{1/2}/K$	0.0	–	–	Temperature dependence of K1
K1SI1	1/K	0.0	–	–	Temperature dependence of K1SI
K1SAT1	$V^{-1/2}/K$	0.0	–	–	Temperature dependence of K1SAT
TMEXP	1/K	0.0	–	–	Temperature coefficient for V_{dseff} smoothing
TMEXPR	1/K	TMEXP	–	–	Reverse-mode temperature coefficient for V_{dseff} smoothing

Name	Unit	Default	Min	Max	Description
PTWGT	1/K	0.004	–	–	PTWG temperature coefficient
PRT	1/K	0.001	–	–	Series resistance temperature coefficient
TRSDR	1/K	0.0	–	–	Source side drift resistance temperature coefficient
TRDDR	1/K	TRSDR	–	–	Drain side drift resistance temperature coefficient
IIT	–	−0.5	–	–	Impact ionization temperature coefficient (IIMOD = 1)
TII	–	0.0	–	–	Impact ionization temperature coefficient (IIMOD = 2)
ALPHA01	mV^{-1}/K	0.0	–	–	Temperature dependence of ALPHA0
ALPHA11	V^{-1}/K	0.0	–	–	Temperature dependence of ALPHA1
ALPHAII01	mV^{-1}/K	0.0	–	–	Temperature dependence of ALPHAII0
ALPHAII11	V^{-1}/K	0.0	–	–	Temperature dependence of ALPHAII1
TGIDL	1/K	−0.003	–	–	GISL/GIDL temperature coefficient
IGT	–	2.5	–	–	Gate current temperature coefficient
AIGBINV1	$(Fs^2/g)^{0.5}m^{-1}/K$	0.0	–	–	Temperature dependence of AIGBINV
AIGBACC1	$(Fs^2/g)^{0.5}m^{-1}/K$	0.0	–	–	Temperature dependence of AIGBACC
AIGC1	$(Fs^2/g)^{0.5}m^{-1}/K$	0.0	–	–	Temperature dependence of AIGC
AIGS1	$(Fs^2/g)^{0.5}m^{-1}/K$	0.0	–	–	Temperature dependence of AIGS
AIGD1	$(Fs^2/g)^{0.5}m^{-1}/K$	0.0	–	–	Temperature dependence of AIGD
TCJ	1/K	0.0	–	–	Temperature coefficient for CJS/CJD
TCJSW	1/K	0.0	–	–	Temperature coefficient for CJSWS/CJSWD
TCJSWG	1/K	0.0	–	–	Temperature coefficient for CJSWGS/CJSWGD
TPB	1/K	0.0	–	–	Temperature coefficient for PBS/PBD

Continued

Name	Unit	Default	Min	Max	Description
TPBSW	1/K	0.0	–	–	Temperature coefficient for PBSWS/PBSWD
TPBSWG	1/K	0.0	–	–	Temperature coefficient for PBSWGS/PBSWGD
XTIS	–	3.0	–	–	Source junction current temperature exponent
XTID	–	XTIS	–	–	Drain junction current temperature exponent
XTSS	–	0.02	–	–	Power dependence of JTSS on temperature
XTSD	–	XTSS	–	–	Power dependence of JTSD on temperature
XTSSWS	–	0.02	–	–	Power dependence of JTSSWS on temperature
XTSSWD	–	XTSSWS	–	–	Power dependence of JTSSWD on temperature
XTSSWGS	–	0.02	–	–	Power dependence of JTSSWGS on temperature
XTSSWGD	–	XTSSWGS	–	–	Power dependence of JTSSWGD on temperature
TNJTS	–	0.0	–	–	Temperature coefficient for NJTS
TNJTSD	–	TNJTS	–	–	Temperature coefficient for NJTSD
TNJTSSW	–	0.0	–	–	Temperature coefficient for NJTSSW
TNJTSSWD	–	TNJTSSW	–	–	Temperature coefficient for NJTSSWD
TNJTSSWG	–	0.0	–	–	Temperature coefficient for NJTSSWG
TNJTSSWGD	–	TNJTSSWG	–	–	Temperature coefficient for NJTSSWGD

Name	Unit	Default	Min	Max	Description
RTH0	$\Omega\,\mathrm{m\,K/W}$	0.01	0.0	–	Thermal resistance for self-heating calculation
CTH0	$\mathrm{W\,s/m/K}$	1.0e−5	0.0	–	Thermal capacitance for self-heating calculation
WTH0	m	0.0	0.0	–	Width-dependence coefficient for self-heating calculation

A.7 PARAMETERS FOR VARIABILITY MODELING

A set of parameters causing variability in device behavior are identified. Users can associate appropriate variability function as appropriate. The list is open to modification with users feedbacks and suggestions. Other than DELVTRAND, UOMULT, and IDS0MULT, the parameters listed here were already introduced previously as either instance parameters or model parameters. All of the following parameters should be elevated to instance parameter status, if required for variability modeling, or should be delegated to a model parameter status (unless introduced before as an instance parameter).

Name	Unit	Default	Min	Max	Description
DTEMP	K	0.0	–	–	Device temperature shift handle
DELVTRAND	V	0.0	–	–	Threshold voltage shift handle
UOMULT	–	1.0	–	–	Multiplier to mobility (or more precisely divides D_{mob}, D_{mobs})
IDS0MULT	–	1.0	–	–	Multiplier to source-drain channel current
TFIN[(i)]	m	15n	1n	–	Body (fin) thickness
FPITCH[(i)]	m	80n	TFIN	–	Fin pitch
XL[(mod)]	m	0	–	–	L offset for channel length due to mask/etch effect
NBODY[(mod)]	$\mathrm{m^{-3}}$	1e22	1e18	5e24	Channel (body) doping concentration
EOT[(mod)]	m	1.0n	0.1n	–	SiO_2 equivalent gate dielectric thickness (including inversion layer thickness)

Name	Unit	Default	Min	Max	Description
TOXP$^{(mod)}$	m	1.2n	0.1n	–	Physical oxide thickness
RSHS$^{(mod)}$	Ω	0.0	0.0	–	Source-side sheet resistance
RSHD$^{(mod)}$	Ω	RSHS	0.0	–	Drain-side sheet resistance
RHOC$^{(mod)}$	Ω m^2	1p	10^{-18}	10^{-9}	Contact resistivity at the silicon/silicide interface
RHORSD$^{(mod)}$	Ω m	Calculated	0	–	Average resistivity of silicon in the raised source/drain region
RHOEXT$^{(mod)}$	Ω m	RHORSD	0	–	Average resistivity of silicon in the fin extension region

Note: Parameters already introduced as instance parameters are marked $^{(i)}$ and model parameters are marked $^{(mod)}$.

Index

Note: Page numbers followed by *f* indicate figures and *t* indicate tables.

A

AC symmetry test, 222–224, 223*f*, 224*f*
Analog compact model
  application, 16–17
  gain-bandwidth product, 16–17
  gain compression, 39–41, 40*f*, 41*f*
  geometric scaling
    DIBL, 21
    performance optimization, 19–21
    three-dimensional structures, 22–23
    threshold voltage *vs.* channel length, 21, 22*f*
  Gummel symmetry test, 42–43, 42*f*, 43*f*
  harmonic distortion
    drain-source voltage, 36
    even harmonics, 37
    G_m nonlinearity, 39
    input sinusoid, 36, 37, 38*f*
    loop gain, 36
    odd harmonics, 37
    power series, 39
    second harmonic *vs.* third harmonic signal, 37–39, 38*f*
    terminal voltage, 36
    transient/steady-state simulation, 39
    weak distortion, 37
  intermodulation distortion, 42
  intrinsic gain
    channel length, 24–25, 25*f*
    Coulomb scattering, 28
    device bias current, 25–26
    gate source *vs.* drain source voltage, 28, 29*f*
    optimum overdrive voltage, 26, 26*f*, 27*f*
    output conductance, 28–31
    output resistance, 28–31, 30*f*
    single-stage amplifier, 24–25
    small-signal parameters, 25
    transconductance efficiency *vs.* gate source voltage, 26, 27*f*
    V_{ds} function, 27–28, 28*f*
  memory effects, 41
  mixed-signal design, 17
  quiescent operation point
    absolute accuracy, 18
    convergence properties, 18
    device current *vs.* absolute bias voltage, 19, 20*f*
    device current *vs.* drain-source voltage, 19, 21*f*
    I-V curves, 17–18
    leakage currents, 19
    Newton-Raphson iteration, 18
    surface potential solution, 18
    threshold voltage, 19
    velocity saturation, 18–19
  signal-dependent charge injection, 17
  speed, unity gain frequency, 31–32, 31*f*, 33*f*
  thermal noise
    common-gate amplifier, 33–35
    extrinsic noise sources, 35
    flicker noise, 35, 35*f*
    intrinsic noise sources, 35
    spot noise variance *vs.* gate bias voltage, 32–33, 34*f*
  variability modeling
    differential circuit, 23–24, 24*f*
    doping and lithography, 24
    physical parameters, 23
    random variations, 23
    systematic variations, 23
    transistor performance, 23
Asymptotic correctness, 216–217

B

Band-to-band tunneling (BTBT) leakage current, 206–207
Body effect model, 119–120
BSIM-CMG model
  bird's-eye view, 156–157, 157*f*, 158, 159*f*
  channel current modeling
    body effect model, 119–120
    channel length modulation, 121–122
    characteristic length model, 100–102
    DIBL effect, 104–105, 122–123
    drain saturation voltage, 109–114
    drain-to-source current, 123
    electrostatic potential profile, 102–104
    geometrical confinement, (*see* Quantum mechanical effect)
    lateral nonuniform doping, 119
    output resistance model, 120–123
    reverse short-channel effect, 105
    SS degradation, 106, 107*f*
    threshold voltage roll-off, 104
    velocity saturation model, 114–118
    vertical field mobility degradation, 109
  cross-sectional diagram, 158, 159*f*

BSIM-CMG model *(Continued)*
  doped double-gate FET
    Boltzmann distribution, 72, 73*f*
    drift-diffusion equation, 79
    fin potential *vs.* fin position, 73–75,
      74*f*, 75*f*
    Gauss's law and boundary condition,
      75–76
    GCA condition, 72, 73*f*
    mobile charge density, 76–77, 78*f*
    parameter COREMOD, 77–79
    perturbation potential, 75–76
    Poisson's equation, 72
    saturation condition, 79–80, 80*f*
    subthreshold region, 79–80, 80*f*
    surface potential, 76–77, 77*f* *(see also*
      Newton-Raphson method)
    triode region, 79–80, 80*f*
  drain charge, 143–144
  flicker noise
    carrier number fluctuation theory, 196
    definition, 196
    drain current, 196
    drain noise spectral density, 197–198
    mobility fluctuation theory, 196
    strong-inversion region, 198
    trap density fluctuation, 197
    unified model, 196
    weak-inversion region, 198
  gate charge, 142–143
  gate electrode resistance model *(see* Parasitic
    resistance)
  gate-induced drain leakage, 130–131
  gate oxide tunneling, 132–133,
    132*f*
  model parameters *(see* Parameter extraction)
  NQS model *(see* Non-quasi-static (NQS) effects)
  parasitic capacitance *(see* Parasitic capacitance)
  quasi-static assumption, 147
  shot noise, 198–199
  SOI, 157–158, 157*f*
  source charge, 144
  source/drain resistance
    components, 159, 160*f*
    contact resistance (R_{con}), 159, 160–161, 161*f*,
      162*f*
    extension resistance (R_{ext}), 159, 165–167,
      165*f*, 166*f*
    spreading resistance (R_{sp}), 159, 161–165,
      162*f*, 164*f*
  symbol definition, 157–158, 158*t*
  thermal noise, 194, 195*f*
  unified compact model

channel doping concentration, 84, 84*f*
Cy-GAA FET, 81–83
drain current normalization, 83–84
long-channel silicon-on-insulator, 85–87,
  86*f*, 87*f*
rounded trapezoidal shape, 80–81, 81*f*
trapezoidal TG FinFET, 84, 85, 85*f*, 86*f*

C
Channel induced gate resistance model, 149–150,
  149*f*
Channel length modulation (CLM), 121–122
Charge/capacitance model
  forward-bias model
    BSIM-CMG, 212–213, 214
    junction capacitance, 211–212, 212*f*, 213
    junction charge, 211–212, 213
    PBS, 211–212
  reverse-bias model
    charge density, 209–210
    junction capacitance, 208, 209–211, 209*f*
    junction depletion depth, 209–210
Charge segmentation model
  boundary conditions, 151
  continuity equation, 150, 152
  gate terminal charge, 153
  spline-collocation method, 150–151
  Ward-Dutton partitioning, 152
CLM. *See* Channel length modulation (CLM)
Compact model
  AC symmetry test, 222–224, 223*f*, 224*f*
  analog metrics *(see* Analog compact model)
  asymptotic correctness, 216–217
  Gummel symmetry test, 220–221, 220*f*, 221*f*
  harmonic balance simulation test, 222, 222*f*, 223*f*
  reciprocity test, capacitances, 225, 225*f*
  RF metrics. *(see* RF compact model)
  self-heating effect, 225–226, 226*f*
  source-drain symmetry, 219–220
  thermal noise, 226
  weak and strong inversion region
    conductance test, 218, 219*f*
    slope ratio test, 217–218, 218*f*
    volume inversion test, 219, 220*f*
Contact resistance (R_{con}), 159, 160–161, 161*f*, 162*f*
Continuous starting function (CSF)
  doping concentration, 89–90, 90*f*, 91*f*, 92*f*, 93*f*
  strong-inversion region, 88–89
  subthreshold region, 88, 89–90
Core model. *See* BSIM-CMG model
CSF. *See* Continuous starting function (CSF)
Cylindrical gate-all-around (Cy-GAA) FET, 81–83

D

Drain-induced barrier lowering (DIBL)
  intrinsic gain, 27–28
  off-state leakage current, 129, 130f
  parameters, 237, 237t
  threshold voltage, 21, 22f, 104–105, 105f,
    122–123
Drain saturation voltage
  RDSMOD = 0, 111–114
  RDSMOD = 1 and 2, 110–111

E

Extension resistance (R_{ext})
  accumulation resistance (R_{acc}), 165–166
  components, 166–167
  definition, 165
  doping profile, 165, 165f
  resistance modeling, 165, 166f
  spacer configurations, 165, 165f

F

Flicker noise
  carrier number fluctuation theory, 196
  definition, 196
  drain current, 196
  drain noise spectral density, 197–198
  mobility fluctuation theory, 196
  phase noise, 63–64, 63f
  standard compact model, 9–10
  strong-inversion region, 198
  thermal noise, 35, 35f
  trap density fluctuation, 197
  unified model, 196
  weak-inversion region, 198

G

Gate-induced drain leakage (GIDL), 9
  BSIM-CMG model, 130–131
  deep depletion, 128–129
  hole layers, 129
  narrow tunneling barrier, 129
  solid-source diffusion, 129
Gate-induced source leakage (GISL), 128, 131
Gate oxide tunneling
  BSIM-CMG model, 132–133, 132f
  gate-to-body, 133–135, 133f, 134f
  gate-to-channel, 135–136, 136f
  high-κ oxide interfaces, 131–132
  source/drain current, 136–137, 137f
Genetic algorithm, 230
GIDL. *See* Gate-induced drain leakage (GIDL)
GISL. *See* Gate-induced source leakage (GISL)

Gradual-channel approximation (GCA) condition,
    72, 73f
Gummel symmetry test
  drain-current model, 24f, 25f, 31–32
  potential symmetry problem, detection, 42–43,
    42f, 43f

H

Harmonic balance simulation test, 222,
    222f, 223f
Hybrid-pi model
  frequency dependence, 44–45
  intrinsic transistor, 44, 44f
  Mason's unilateral gain U, 54–55, 55f
  maximum gain
    capacitance neutralization, 53–54, 53f
    channel resistance, 51, 52f
    physical gate resistance, 51, 52f
    power gain, 51, 51f
    quadrature phase relation, 49–50, 49f
    source and load admittance, 48–49
    stability factor, 48–49, 50, 51f
  pi-circuit parameters, 45–46, 45f, 47f
  small-signal assumption, 44–45
  tuned amplifier, 46–48, 48f
  Y parameters, 45–46, 45f

I

Impact ionization model
  intrinsic gain, 28f, 134
  temperature dependence, 249
  velocity saturation model, 138–139
Induced gate noise, 194, 195f
Inversion charge density, 142, 176

J

Johnson–Nyquist noise. *See* Thermal noise model
Junction diode
  current model
    BSIM-CMG model, 204, 205f, 206, 207–208
    BTBT-based current, 206–207
    first-order Taylor expansion, 205, 206
    ideal junction current, 203
    quadratic equation, 204
    reverse-bias saturation current, 203–204
    SRH leakage current, 206, 207–208
    TAT leakage current, 207–208
  junction capacitance (*see* Charge/capacitance
    model)
  planar bulk MOSFETs, 201–203
  p-type FinFET, 201–203, 202f
  punch-through stop implant, 201–203, 202f

L

Lateral nonuniform doping model, 119
Leakage current modeling. *See* Off-state leakage current
Levenberg-Marquardt algorithm, 230

M

Metal-oxide-semiconductor field-effect transistor (MOSFET)
  charge segmentation, 150*f*
  drain current, 196
  lithography scaling
    germanium and group III-V materials, 5
    manufacturing cost, 5
    shallow trench isolation oxide, 4, 5*f*
    subthreshold swing, 5, 6*f*
    thin body transistor, 4, 5*f*
  power dissipation, 125
  saturation region, 121, 121*f*, 122*f*
  short-channel effects
    gate critical dimension variation, 2, 2*f*
    gate oxide thickness, reduction, 2–3, 2*f*, 3*f*
    leakage paths, 3, 3*f*
    random dopant fluctuation, 2, 2*f*
  standard compact model
    accuracy, 9, 10*f*
    BSIM-CMG, 8–9, 11–12
    CMOS technology, 9
    core model, 9
    flicker noise, 9–10
    germanium, 10, 11*f*
    GIDL, 9
    random telegraphic noise, 9–10
    real device model, 9, 10, 10*f*
    SPICE simulation, 7–8, 8*f*
    transfer characteristics, 10, 11*f*
  ultrathin body
    advantages, 6
    channel doping, 4, 4*f*
    leakage current, simulation, 6–7, 7*f*
    nanometers, 6–7, 6*f*
    random dopant fluctuation, 4, 4*f*
    temperature bias instability, 4
    tunneling leakage, 4
    two-dimensional semiconductors, 7
  weak-inversion current, 127–128
Mobility degradation, 109

N

Newton-Raphson method
  continuous starting function
    doping concentration, 89–90, 90*f*, 91*f*, 92*f*, 93*f*

strong-inversion region, 88–89
subthreshold region, 88, 89–90
  quartic modified iteration
    explicit surface potential model, 91–95, 95*f*, 96*f*
    high-order correction, 91–95
Non-quasi-static (NQS) effects
  channel induced gate resistance model, 149–150, 149*f*
  charge segmentation model
    boundary conditions, 151
    continuity equation, 150, 152
    gate terminal charge, 153
    spline-collocation method, 150–151
    Ward-Dutton partitioning, 152
  effective transconductance G_m vs. frequency, 56, 58*f*
  lumped circuits, 56–57, 58*f*
  RC time constant, 55–56
  relaxation time approximation model, 147–149, 148*f*
  $Y11$ component, 56, 57*f*
NQS effects. *See* Non-quasi-static (NQS) effects

O

Off-state leakage current
  gate-induced drain leakage
    BSIM-CMG model, 130–131
    deep depletion, 128–129
    hole layer, formation, 129
    narrow tunneling barrier, 129
    solid-source diffusion, 129
  gate oxide tunneling
    BSIM-CMG model, 132–133, 132*f*
    gate-to-body, 133–135, 133*f*, 134*f*
    gate-to-channel, 135–136, 136*f*
    high-κ oxide interfaces, 131–132
    source/drain current, 136–137, 137*f*
  impact ionization model, 138–139
  weak-inversion current
    electrostatic control, 128
    source-channel barrier, 127, 127*f*
    subthreshold swing, 126*f*, 127

P

Parameter extraction
  core model derivation, 229
  genetic algorithm, 230
  global parameters, 230
    DIBL effect parameters, 237, 237*t*
    drain-current model parameters, 239
    extraction flowchart, 231, 231*t*, 232*f*

interface charge, 235, 235t
low-field-mobility, 235, 236, 236f, 236t
mobility parameters, 235, 235t
resistance scaling, 232, 232f
saturation I_d- V_g characteristics, 238, 239f
subthreshold slope parameters, 234, 234f
threshold voltage difference, 233, 233f
velocity saturation, 237, 238, 238t
work function, 235, 235t
Levenberg-Marquardt algorithm, 230
local parameters, 230
particle swarm optimization, 230
real-device effects, 229, 240f
Parasitic capacitance
three-dimensional fringe capacitance, 185–186, 185f, 186f
two-dimensional fringe capacitance
contact-to-gate capacitance (C_{cg}), 180f, 182, 183, 184
fin-to-gate capacitance (C_{fg}), 180, 180f, 181f
Parasitic resistance
contact resistance model, 177
definition, 155
device optimization, 168–170, 169f
extension resistance model, 176
gate electrode resistance model, 177
gate resistance, 156
individual resistance components, 174, 175f
NQS, 177, 178f
physical parameters, 176
SEG process, 156
source/drain resistance
effective channel length, 171–172, 172f
extracted model parameters, 173, 174t
gate voltage, 171–173, 172f, 173f
potential issues, 170
spacer thickness, 173, 174f
total channel resistance, 170–171, 171f
$vs.$ raised source/drain length, 174, 175f
spreading resistance model, 176
TCAD simulation (see TCAD simulation)

Q

Quantum mechanical effect
charge centroid, 115–116, 116f, 117f, 119
effective oxide thickness/capacitance, 118
effective width model, 118
threshold voltage, 107–108, 108f

R

Relaxation time approximation model, 147–149, 148f
Reverse short-channel effect, 105

RF compact model
intermodulation distortion
second-order intermodulation, 64, 65, 66f
spectral regrowth, 67–68
third-order intermodulation, 64–65, 66f
two-tone signal, 64
Volterra series, 67
minimum achievable noise figure, 59, 60f
noise voltage, 58–59, 58f
NQS
effective transconductance G_m $vs.$ frequency, 56, 58f
lumped circuits, 56–57, 58f
RC time constant, 55–56
$Y11$ component, 56, 57f
phase noise
digital communication system, 62–63, 62f
flicker noise, 63–64, 63f
frequency domain, 61, 61f
local oscillator, 61–62, 62f
Lorentzian spectrum, 63
Pospiesalski noise model, 59–61
two-port parameters (see Hybrid-pi model)

S

Selective epitaxial growth (SEG) process, 156–157
Self-heating effects, 225–226, 226f
Shockley-Reed-Hall (SRH) current, 206, 207–208
Short-channel effects (SCE)
channel length modulation, 121–122
characteristic length model, 100–102
DIBL effect, 104–105, 122–123
gate critical dimension variation, 2, 2f
gate oxide thickness, reduction, 2–3, 2f, 3f
geometrical confinement (see Quantum mechanical effect)
leakage paths, 3, 3f
output resistance model, 120–123
random dopant fluctuation, 2, 2f
reverse short-channel effect, 105
SS degradation, 106, 107f
threshold voltage roll-off, 104
velocity saturation model, 114–118
vertical field mobility degradation, 109
Shot noise, 198–199
Silicon-on-insulator (SOI)
drain current $vs.$ gate voltage, 85–87, 86f
low thermal conductivity, 254
measured data validation, 255–258, 256f, 257f
simulation, 6–7, 7f
source-to-drain direction, 157–158, 157f
SS model, 75f, 163
Slope ratio test, 217–218, 218f

Spreading resistance (R_{sp})
  cross-sectional area, 162
  resistance, 163
  slope factor, 164*f*, 165
  source/drain region, 161–162, 162*f*
  test structure, 164–165, 164*f*
Subthreshold slope (SS) degradation,
    106, 107*f*

T
TCAD simulation
  bird-eye view, 167, 168*f*
  end-fin cases, 188, 189*f*
  fin-to-gate capacitance, 186–187, 187*f*
  four-fin FinFET, 188, 189*f*
  middle-fin cases, 188, 188*f*
  parameters, 167, 169*t*
  source/drain to gate capacitance, 186–187,
    187*f*, 188*f*
Temperature dependence
  leakage current
    gate current, 248
    gate-induced drain/source leakage, 249
    impact ionization, 249
  measured data validation, 255–258, 256*f*, 257*f*
  mobility, 247
  nonsaturation effect, 248
  parasitic source/drain resistances, 249–250
  saturation velocity, 247–248
  self-heating effect, 254–255, 254*f*, 255*f*
  semiconductor properties
    band gap, 244
    density of states, 245
    intrinsic carrier concentration, 245
    source-drain built-in potential, 245
  source/drain diode characteristics
    direct current model, 250–252
    junction capacitances, 252
    trap-assisted tunneling saturation current,
      252–254
  threshold voltage
    body effect, 246
    drain induced barrier lowering, 246
    subthreshold swing, 246–247
  validation range, 255
Thermal noise model
  charge-based model, 194
  common-gate amplifier, 33–35

extrinsic noise sources, 35
  flicker noise, 35, 35*f*
  induced gate noise, 194, 195*f*
  intrinsic noise sources, 35
  noise spectral density, 194
  parameter NTNOI, 194
  saturation and linear region, 194, 195*f*
  spot noise variance *vs.* gate bias voltage, 32–33,
    34*f*
  tests, 226
Threshold voltage roll-off, 104, 105*f*
Transcapacitance, 144–145, 145*f*, 146*f*
Trap-assisted tunneling (TAT) leakage current,
    207–208

U
Ultrathin body (UTB) silicon
  advantages, 6
  channel doping, 4, 4*f*
  leakage current, simulation, 6–7, 7*f*
  nanometers, 6–7, 6*f*
  random dopant fluctuation, 4, 4*f*
  temperature bias instability, 4
  tunneling leakage, 4
  two-dimensional semiconductors, 7
Unified compact model
  channel doping concentration, 84, 84*f*
  Cy-GAA FET, 81–83
  drain current normalization, 83–84
  long-channel silicon-on-insulator, 85–87,
    86*f*, 87*f*
  rounded trapezoidal shape, 80–81, 81*f*
  trapezoidal TG FinFET, 84, 85,
    85*f*, 86*f*

V
Velocity saturation model
  channel current modeling, 114–118
  impact ionization model, 138–139
  temperature dependence, 247–248
Volume inversion test, 219, 220*f*

W
Ward-Dutton partitioning, 143–144, 145,
    146*f*, 152
Wentzel-Kramers-Brillouin (WKB) approximation,
    130

Printed in the United States
By Bookmasters